Laser-Aided Diagnostics of Plasmas and Gases

Series in Plasma Physics

Series Editors:

Professor Peter Stott, CEA Caderache, France
Professor Hans Wilhelmsson, Chalmers University of Technology, Sweden

Other books in the series

An Introduction to Alfvén Waves
R Cross

Transport and Structural Formation in Plasmas
K Itoh, S-I Itoh and A Fukuyama

Tokamak Plasma: a Complex Physical System
B B Kadomtsev

Electromagnetic Instabilities in Inhomogeneous Plasma
A B Mikhailovskii

Instabilities in a Confined Plasma
A B Mikhailovskii

Physics of Intense Beams in Plasma
M V Nezlin

The Plasma Boundary of Magnetic Fusion Devices
P C Stangeby

Collective Modes in Inhomogeneous Plasma
J Weiland

Forthcoming titles in the series

Plasma Physics via Computer Simulation, 2nd Edition
C K Birdsall and A B Langdon

The Theory of Photon Acceleration
J T Mendonça

Nonlinear Instabilities in Plasmas and Hydrodynamics
S S Moiseev, V G Pungin, and V N Oraevsky

Inertial Confinement Fusion
S Pfalzner

Introduction to Dusty Plasma Physics
P K Shukla and N Rao

Series in Plasma Physics

Laser-Aided Diagnostics of Plasmas and Gases

Katsunori Muraoka

Kyushu University

Mitsuo Maeda

Kyushu University

Institute of Physics Publishing
Bristol and Philadelphia

© IOP Publishing Ltd 2001

All rights reserved. No part of this publication may be reproduced, stored in a retrieval system or transmitted in any form or by any means, electronic, mechanical, photocopying, recording or otherwise, without the prior permission of the publisher. Multiple copying is permitted in accordance with the terms of licences issued by the Copyright Licensing Agency under the terms of its agreement with the Committee of Vice-Chancellors and Principals.

British Library Cataloguing-in-Publication Data

A catalogue record for this book is available from the British Library.

ISBN 0 7503 0643 2

Library of Congress Cataloguing-in-Publication Data are available

Consultant Editors: P Stott and H Wilhelmsson
English translation by Mark Bowden 1999

Commissioning Editor: John Navas
Production Editor: Simon Laurenson
Production Control: Sarah Plenty
Cover Design: Frederique Swist
Marketing Executive: Colin Fenton

Published by Institute of Physics Publishing, wholly owned by The Institute of Physics, London

Institute of Physics Publishing, Dirac House, Temple Back, Bristol BS1 6BE, UK

US Office: Institute of Physics Publishing, The Public Ledger Building, Suite 1035, 150 South Independence Mall West, Philadelphia, PA 19106, USA

Typeset by Mackreth Media Services, Hemel Hempstead, Herts
Printed in the UK by Bookcraft, Midsomer Norton, Somerset

Contents

QC
718
.5
D5
M8613
2001
PHYS

Foreword			ix
PART I	**FUNDAMENTALS**		**1**
1	**Laser-Aided Diagnostics of Gases and Plasmas**		**3**
	1.1	Properties of Gases	4
		1.1.1 Classification of Gaseous States	4
		1.1.2 Fundamental Parameters used to describe Gases	5
	1.2	Properties of Plasmas	6
		1.2.1 Different Areas of Plasma Applications	6
		1.2.2 Fundamental Parameters used to describe Plasmas	8
		1.2.3 Summary	11
	1.3	Different States of Matter and their Kinetic Properties	13
	1.4	Characteristics of Laser Light	14
		1.4.1 Coherence	14
		1.4.2 Short Pulse Generation	19
	1.5 Advantages of Laser-Aided Measurement Methods		21
	References		25
2	**Basic Principles of Different Laser-Aided Measurement Techniques**		**26**
	2.1	Interaction of Electromagnetic Waves with Single Particles	27
		2.1.1 Thomson Scattering by Charged Particles	28
		2.1.2 Mie and Rayleigh Scattering	29
		2.1.3 Raman Scattering	31
		2.1.4 Resonant Absorption	32
		2.1.5 Photo-Ionization	33
	2.2	Laser Propagation through Gases and Plasmas	33
		2.2.1 Reflection	33
		2.2.2 Transmission	36
		2.2.3 Refraction	43
		2.2.4 Scattering	45
		2.2.5 Photo-Ionization	52
	2.3	Spectral Profile Measurements	54

v

	2.3.1	Summary of Line Broadening Mechanisms	54
	2.3.2	Examples of Spectral Widths	56
	2.3.3	Spectral Profile Measurement Techniques	57
References			59

3 Hardware for Laser Measurements — 60
3.1 Lasers — 60
    3.1.1 Overview of Laser Systems — 60
    3.1.2 Control of Laser Light — 66
    3.1.3 Gas Lasers — 70
    3.1.4 Solid-State and Semiconductor Diode Lasers — 75
    3.1.5 Tunable Lasers — 79
3.2 Nonlinear Wavelength Conversion Devices — 82
    3.2.1 Nonlinear Optical Effects — 82
    3.2.2 Higher Harmonic Generation and Frequency Mixing — 85
    3.2.3 Optical Parametric Oscillators — 87
    3.2.4 Stimulated Scattering — 88
3.3 Optical Elements and Optical Instruments — 90
    3.3.1 Dispersion Elements and Spectrometers — 90
    3.3.2 Interferometers — 93
    3.3.3 Optical Waveguides — 97
    3.3.4 Other Optical Elements — 99
3.4 Detectors and Signal Processing — 103
    3.4.1 Light Detectors — 103
    3.4.2 Imaging Detectors — 108
    3.4.3 Noise Sources and Signal Recovery — 110
    3.4.4 Observation of Fast Waveforms — 112
References — 113

PART II APPLICATIONS AND MEASUREMENTS — 115

4 Plasma Measurements — 117
4.1 Overview of Plasma Spectroscopic Methods — 117
4.2 Laser-Aided Measurements in High-Temperature Plasmas — 121
    4.2.1 Measurement of Plasma Density and Temperature — 121
    4.2.2 Measurement of Density and Temperature of Neutral and Impurity Species — 131
    4.2.3 Measurement of Electric and Magnetic Fields and Plasma Fluctuations — 136
4.3 Laser-Aided Measurements in Discharge Plasmas — 138
    4.3.1 Measurement of Electric Field — 139
    4.3.2 Measurement of Electron Density and Temperature — 153
    4.3.3 Measurement of Reaction Products — 179
References — 185

5	**Combustion Measurements**		**188**
	5.1 Combustion Fields and Laser-Aided Measurements		188
		5.1.1 Measurement of Particle Densities	189
		5.1.2 Measurement of Temperature	193
		5.1.3 Measurement of Velocity	196
	5.2 Examples of Combustion Measurements		196
		5.2.1 Measurements by Laser-Induced Fluorescence Spectroscopy	197
		5.2.2 Measurements by Coherent Anti-Stokes Raman Spectroscopy	201
		5.2.3 Measurements by Degenerate Four-Wave Mixing	204
	References		206
6	**Measurements in Gas Flow Systems**		**207**
	6.1 Measurement of Refractive Index Changes (Density Measurements)		208
		6.1.1 Schlieren Method	209
		6.1.2 Shadowgraphy	211
		6.1.3 Interferometry	212
		6.1.4 Holography	213
	6.2 Measurement of Flow Velocity		214
		6.2.1 Measurement Techniques	215
		6.2.2 Examples of Measurements	218
	6.3 Imaging of Gas Flows by Laser-Induced Fluorescence		222
		6.3.1 Measurement of Density Distributions	224
		6.3.2 Measurement of Temperature Distributions	226
	References		229
7	**Laser Processing Measurements**		**230**
	7.1 Laser Processing		230
	7.2 Measurement Methods in Laser Processing		234
		7.2.1 Different Methods and their Advantages	234
		7.2.2 Detection of Atomic and Molecular Species	237
	7.3 Examples of Laser Processing Measurements		240
		7.3.1 Measurements of Laser CVD Processes	241
		7.3.2 Measurements of Laser Ablation Processes	247
	References		256
8	**Analytical Chemistry**		**258**
	8.1 Analytical Chemistry and Laser Spectroscopy		258
	8.2 Examples of Analysis using Laser Spectroscopic Techniques		260
		8.2.1 Analysis using Laser Raman Spectroscopy	260
		8.2.2 Analysis using Laser-Induced Emission Spectroscopy	262
		8.2.3 Analysis using Laser-Induced Fluorescence Spectroscopy	264

	8.2.4 Analysis using Laser Ionization Spectroscopy	268
	8.2.5 Analysis using Laser Photothermal Spectroscopy	271
References		274

9 Remote Sensing — 276

9.1 LIDAR and Monitoring of the Atmosphere — 276
    9.1.1 LIDAR Theory — 277
    9.1.2 Different LIDAR Techniques — 278
9.2 Representative LIDAR Experiments — 280
    9.2.1 Mie Scattering LIDAR — 280
    9.2.2 Rayleigh Scattering LIDAR — 283
    9.2.3 Differential Absorption LIDAR (DIAL) — 285
    9.2.4 Raman LIDAR — 289
References — 290

Index — 291

Foreword

The most important property of laser-aided diagnostic techniques is their *active* nature. In the most general laser-aided measurement, an object is irradiated with a laser and the response to the laser light is observed. The light from laser sources can have high coherence, and properties such as intensity, wavelength, pulse length and spectral profile can be controlled. These properties allow measurements to be made with accuracy, sensitivity, and spatio-temporal resolution that is not possible with other methods.

In high-temperature gases, such as combustion gases, chemically active gases and plasmas, intrusive measurement techniques, such as probe methods, are difficult to use. Laser-aided measurement methods, however, with their remote sensing capability, are well suited for measurements in these circumstances. The different responses to laser irradiation can be used to determine not only the kinds of species present and their densities but also a variety of information including velocity, temperature, electric and magnetic fields, and the size and density of macro-particles. By exploiting the directionality of the laser beam, measurements can be made even from objects that are located many kilometres away. This remote sensing capability is especially useful for monitoring the Earth's atmosphere.

Laser-aided measurement techniques are used in almost all fields of science and technology. This book is aimed specifically at people who are interested in applying these techniques to probe gaseous media by providing both descriptions of the basic principles of the main measurement techniques and examples of actual measurements in many research fields. In the past, researchers interested in laser-aided measurement methods have had to search among textbooks and research journals in many different fields, such as combustion, gas dynamics, environmental monitoring, analytical chemistry, plasma physics, optics and laser engineering. There has long been a need for a self-contained book that describes the common aspects of this multi-disciplinary field. We hope that this book goes some part of the way toward meeting that need.

This book is structured in the following way. Part I: Fundamentals covers the basic areas of gases and plasmas, and the laser and detection hardware that is needed to measure them. Part II: Applications and Measurements contains descriptions of specific experiments in each of the research fields mentioned above. We expect that most readers would start in Part I and then go on to Part II, but it is possible to go directly to any particular chapter in Part II and then refer back to Part I when necessary.

We hope that this book is read widely so that the extremely useful laser-aided measurement techniques described here can become used by a wider range of people. This, in turn, should generate new measurement needs that stimulate further development of existing methods, and fresh development of new methods. This would allow the field of laser-aided diagnostics to flourish even further in the future.

This book is a translation of a Japanese book published in 1995. The original text has been revised to take into account the many developments that have occurred in this field during the last five years. In writing this book, we benefited greatly from being able to refer to the research of many different groups, and the original research articles are cited throughout the text. Without the cooperation of the authors of these articles, this kind of review book would not be possible. In particular, the research reported from our own groups was performed in collaboration with our colleagues at Kyushu University: Drs C Honda, T Okada, K Uchino, M Uchiumi, T Kajiwara and M D Bowden. We wish to express our deep appreciation to these people. In addition, the translation of the original Japanese text was performed by Dr M D Bowden, and we thank him for his hard work.

The idea of publishing this book in the present English translation was proposed by Dr Peter Stott, a lifelong friend of one of the authors (KM). His initiative and continuous support during the preparation of the English version is greatly appreciated, for without this, this book would not have been possible.

K Muraoka, M Maeda
March 2000

PART I

FUNDAMENTALS

Chapter 1

Laser-Aided Diagnostics of Gases and Plasmas

This chapter is an overview of some of the important features of gases and plasmas, and of the role that laser-aided diagnostics play in understanding them. Before going on to the main parts of this chapter though, we will briefly summarize the importance of gases and plasmas in society, and in particular, in the industrial sector.

The different types of matter that exist in the universe often are classified into the so-called three states of matter—solids, liquids and gases. From the viewpoint of the mechanical properties of each type of material however, the first two categories can be combined into what we might call *condensed matter*, with gases being a quite separate category that we can refer to as *non-condensed matter*. In condensed matter, the distance between the atoms is of the same order as the size of the atoms themselves ($\sim 10^{-10}$ m), and this results in the potential energy between the particles being much larger than the kinetic energy of the individual particles. In non-condensed matter, however, the energy relationship is reversed, and the particles can move freely in the space that exists between them. This has the consequence that, while condensed material has virtually no response to changes in pressure, the properties of gases are affected very much by pressure.

Condensed matter consists of a great number of particles packed closely together, and the many different types of condensed matter possess a wide variety of mechanical and electrical characteristics that are exploited in different scientific and technological applications. These materials and their applications are recognized widely by virtually everyone in society.

Until recently, however, most people rarely were conscious of the existence of gases, even though when viewed on the scale of the universe, gases comprise more than 99% of all matter. Even in industry, interest was limited mainly to combustion processes, used to generate energy, and gas dynamics, used in the aerospace industry. In recent years, though, immediate problems such as pollution, and potential problems such as global warming and ozone depletion, have made society more aware of the importance of the atmosphere and the gases in it. In addition, the state of matter called plasma, created when atoms or molecules in a gas are ionized and separate into electrons and ions, has begun to play a significant role in many industries. Electrons and ions have much greater chemical activity than neutral particles, and applications of plasmas, based on the various physical, chemical and nuclear reactions that these particles can induce, have become widespread.

In other words, we are now at a time when the science and technology that is supported by gases and plasmas has become a central part of our lives. Non-condensed matter has an equal, and in some cases greater, importance to condensed matter.

As mentioned earlier, this chapter is a summary of general features of gases, plasmas and laser diagnostics. In the sections below, the parameters that are used to characterize gases and plasmas are described, with the emphasis on those characteristics that are important for understanding laser-aided diagnostic measurements.

1.1 Properties of Gases

Section 1.1.1 contains a description of different types of gaseous media and related applications. In section 1.1.2, the basic parameters that are used to describe gases are summarized.

1.1.1 Classification of Gaseous States

The different gaseous media that either exist naturally or are used in industrial processes can be classified in the four broad areas described below.

The atmosphere. The Earth's atmosphere contains a variety of different gases. The main gases are nitrogen (N_2, molecular weight 28, volume concentration 78.1%, and mass concentration 75.5%) and oxygen (O_2, molecular weight 32, volume concentration 20.9%, and mass concentration 23.1%), but argon (Ar, mass concentration 1.3%), carbon dioxide (CO_2) and carbon monoxide (CO) also exist in significant quantities. In addition to these atoms and molecules, various kinds of aerosols (small liquid or solid particles with radii of the order of 1 μm), water vapour and ozone are present, together with small quantities of other species. These trace species, together with gases such as CO_2, NO_x and SO_x, are related to phenomena such as air pollution, global warming and ozone layer depletion that have become major concerns in recent years. For an understanding of these phenomena, it is necessary to be able to determine the distributions of these species, with respect to both location and altitude, as well as measuring the change of these distributions with time.

Detection of these species in the atmosphere using LIDAR (light detection and ranging) is the main subject of chapter 9.

Gases in fans and compressors. Fans and compressors are machines that are used to transport gases and compress them to high pressure. The main difference between them is that for fans, the ratio of pressure before and after the machine is low, while for compressors, this ratio is high. The kind of fan used in homes is a common example of this type of device, but fans and compressors are used in chemical plants, car and aeroplane engines, and in nearly all types of industries. Many different gases are used in the devices, including air, different types of pure gases, and mixtures of gases, liquids and solids. In these kinds of devices, the aim is to efficiently convert electrical energy to gas kinetic energy while maintaining the desired compression ratio or particle flux. Part of the research that is necessary to improve the efficiency of fans and compressors involves measurement of gas properties both inside and around the machine itself. Information about quantities such as flow speed, particle density, temperature and pressure, and the fluctuations in these quantities, is desirable.

Chapter 6 covers the topic of laser-aided diagnostics of these quantities.

Gases inside vacuum equipment and pressure equipment. Any device in which the atmosphere is sealed off and a vacuum pump is used to reduce the internal pressure can be considered broadly to be a piece of vacuum equipment. Similarly, any device in which the internal pressure is raised above atmospheric pressure can be termed pressure equipment. These kinds of equipment are widely used in industrial environments such as chemical factories, semiconductor fabrication facilities and food processing plants. For the optimization of each device, knowledge of the gas composition, and the density, velocity and temperature of each species is usually necessary.

The techniques described in chapters 6 and 8 are suitable for these measurements.

Combustion gases. Combustion is the process in which a fuel reacts with oxygen in the atmosphere to burn, releasing light and heat energy. The energy produced by the combustion process, whether used in the form of heat, as in a boiler or a heat source, or utilized in the form of the resulting gas expansion, as in an internal combustion engine, is one of the most basic elements of virtually all modern industrial activity. Most fuels are hydrocarbon materials refined from coal or oil. CO_2, CO, H_2O and OH are the principal gas species produced by combustion. Some other by-products of combustion, such as NO_X, SO_X and dioxin, are associated with recent environmental problems.

Chapter 5 covers the topic of laser-aided diagnostics of these quantities in combustion environments.

1.1.2 Fundamental Parameters used to describe Gases

As mentioned at the beginning of this chapter, amongst the types of substances that comprise the world around us, gases differ from solids and liquids because the space between the particles in a gas is extremely large compared with the size of the particles themselves. To express this in a different way, the kinetic energy of the particles is much greater than the potential energy between the particles. Hence, except for when they collide with each other or with walls, gas particles move in straight lines. The basic parameters that describe gases are parameters that describe the number of particles for each constituent species, and parameters that describe the particle motion. For the main species, those with concentrations of more than a few per cent, the number of particles is specified by either the density ρ_j, which has units of kg m^{-3} ($j = 1, 2, \ldots$ refers to each individual species) or by the number density n_j, which has units of m^{-3} and is defined as ρ_j/m_j, where m_j is the mass of the jth species. In addition to the main constituent species, there may be other species with concentrations of the order of ppm (10^{-6}) or ppb (10^{-9}), and there also may exist small particles with size of the order of 1 μm suspended in the medium. Although these may be present only in minute quantities, the trace species sometimes have important effects. Climate and pollution problems are good examples of such effects.

The most important parameters related to particle transport and motion are the particle temperatures T_j (and the partial pressures p_j, calculated using $p_j = n_j k T_j$, where k is Boltzmann's constant) and the particle velocities v_j. In addition, each quantity X fluctuates around its average value $\overline{X}$, and the fluctuation $x = X - \overline{X}$ sometimes can be important. For example, fluctuations are important when considering the friction of moving parts inside a device such as a fan or a compressor.

When collisions between particles become rare, as is the case in vacuum devices, the continuum approach on which the above definitions of density and temperature are based becomes inadequate. In these circumstances, gas kinetic theory becomes necessary. This theory describes the motion of individual particles, and contains information of the velocity and energy distribution functions, and the conditions at the boundary, such as at the walls where reflection, absorption and emission occur.

1.2 Properties of Plasmas

A plasma is a gas in which molecules and atoms have been ionized to a considerable degree and for this reason a plasma often is referred to as an ionized gas. Section 1.2.1 contains a general overview of plasma applications, and section 1.2.2 is a summary of the fundamental aspects of plasmas that are necessary to understand the following chapters. Section 1.2.3 contains an overview of current areas of plasma research.

The topic of *laser-aided plasma diagnostics* (LAPD) is described in detail in chapter 4. Chapter 7 covers the somewhat related topic of diagnostics of laser processing applications.

1.2.1 Different Areas of Plasma Applications

The uses of plasmas can be classified into several areas according to the area of application. There are applications that use the thermal and chemical energy of plasmas, applications that use the kinetic energy of plasmas, applications that use the electromagnetic waves (and in particular the light) radiated from plasmas, and thermonuclear fusion. These different areas are outlined briefly below.

Applications that use the thermal and chemical energy of plasmas. A plasma is at a much higher energy state than ordinary matter, and the thermal energy or the electron activity that results from that energy can be used to melt materials, generate new particle species, and initiate a variety of different kinds of processes. Different application areas, together with some specific examples, are listed below:

- **arc plasma processing:** arc welding, plasma jet cutting, steel refinery processes;
- **material fabrication:** single crystal growth, fabrication of small particles and powders;
- **plasma chemistry:** synthesis of acetylene, cyanide, oxygen and nitrogen-containing compounds;
- **semiconductor processing:** etching, deposition, formation of thin films of amorphous silicon and superconductors;
- **surface modification:** control of hydrophobicity/hydrophilicity and surface micro-roughness etc.

Applications that use the kinetic energy of plasmas. Because plasmas possess electrical conductivity, they can be affected by electric and magnetic fields. Electrical energy can be extracted from the plasma kinetic energy, and vice versa. Applications of this type include the following:

- **magnetohydrodynamic (MHD) power generation:** coal, oil or natural gas combustion products with alkali metal addition, closed cycle MHD using nuclear power as a heat source;
- **propulsion in space:** control of the position and orientation of spacecraft.

Applications that use the electromagnetic waves radiated from plasmas. Plasmas interact with electromagnetic waves, and in particular light, in many ways. Some applications that use this aspects of plasmas are listed below:

- **light sources:** lighting, illumination, light sources for copying, backlighting for crystal displays;
- **laser pumping:** gas laser discharges, flashlamp excitation for solid-state and liquid-state lasers;
- **plasma display panels:** high-definition televisions, display panels for industrial machinery, large-scale road signs.

Controlled thermonuclear fusion. All of the Earth's energy sources are generated directly or indirectly by the sun. In 1938, Baethe and Weizsaecker proposed that the sun's energy comes from fusion of light elements. Further research then established that the energy in all stars came from fusion reactions. Research aimed at 'producing the energy of a star on Earth' began in the second half of the 1940s and the first realization of this came in the rather unhappy form of the hydrogen bomb. Since that time, the research aim has changed to the generation of *controlled* thermonuclear fusion for peaceful purposes.

There has been a large-scale effort to develop controlled thermonuclear fusion as the basis of a new energy source. Initial problems of plasma configuration largely have been overcome by developments in plasma theory and technology, and this ultimate energy source for the human race is expected to be realized within a generation.

1.2.2 Fundamental Parameters used to describe Plasmas

As for a gas, the basic parameters that are necessary to describe a plasma state are those concerning particle densities and particle motion. Compared with a gas, however, a plasma is a much more complicated medium, and various important characteristic lengths and characteristic frequencies exist. The basic theory of plasma physics and plasma engineering can be found in many textbooks on these subjects [1,2]. This section contains a summary of only those basic parameters that are necessary for the understanding of laser-aided measurements of plasmas.

Density. The density of each species in a plasma is the first important parameter, and among these, the electron and ion densities are the most important. The number density n_j ($j = e,i$) rather than the mass density ρ_j ($\equiv m_j n_j$) is more commonly used, and this quantity is frequently called simply the density. In some plasmas, there are multiple-ionized ions, and the density of each ionized state may be important. There are also cases when negative ions, formed by the attachment of electrons to atoms and molecules, can be important in determining plasma behaviour.

The difference between ion and electron densities gives the space charge $\sigma = e(n_i - n_e)$ (the ions are assumed to be singly ionized). Another important parameter is the plasma current density $j = (n_i v_i - n_e v_e)$, where v_i and v_e are the average ion and electron velocities respectively. The electric and magnetic fields in a plasma can be determined, in principle, by using σ and j as source terms in Maxwell's equations, with given boundary conditions.

Another important property is the density of neutral particles. The plasma is generated and maintained by collisions of electrons with neutral species in the background gas. In these collisions, ionization and dissociation can lead to the formation of new particle species. Unstable neutral particles, called *radicals*, can be produced by these collisions, and the radical species often are extremely important in plasma processing applications. In other areas, too, there are many cases in which knowledge of the densities of these particles is necessary for understanding of basic mechanisms and applications of plasmas.

For some plasma conditions, small particles and/or powders that consist of large numbers of individual atoms and molecules are generated in the plasma. These particles can be a problem in some plasma applications. For example, in plasma-aided deposition and etching of thin films, particles with size of up to 1 μm sometimes are produced in the discharge, and these particles become embedded in the film surface, seriously affecting the film quality. Also, in high-temperature plasmas, small particles sometimes are produced when the plasma comes into contact with the walls, or other surfaces, and the presence of these particles results in a deterioration of the plasma performance. In these situations, it is necessary to know the density and size distribution of the particles.

Temperature and velocity (energy) distribution function. The most important parameter related to particle transport and motion is the temperature, but it should be noted that there are many plasmas in which there is a large difference between the temperatures of the electrons and ions, and a large difference in their response to electromagnetic fields. This is due to the fact that the ions and electrons are not in energy equilibrium in many plasma conditions. This can be understood by considering the equation of motion for the charged particle species in an electromagnetic field

$$m_j \frac{d\mathbf{v}_j}{dt} = q_j (\mathbf{E} + \mathbf{v}_j \times \mathbf{B}) \quad (1.1)$$

where $j = e, i$. The electron acceleration due to the fields is much larger than the ion acceleration, because $m_e \ll m_i$ and $q_e \sim q_i$. In addition, although the rate of energy transfer is large for the case of electron–electron collisions and ion–ion collisions, the energy exchange in an electron–ion collision is of the order of m_e/m_i. Because of this, there are many plasmas in which the electron and the ion groups are at different thermal equilibria. In these cases, two temperatures, called the *electron temperature* and the *ion temperature*, are defined separately.

If the number of collisions inside the electron or ion groups is insufficient for thermal equilibrium in the group to be established, the velocity distribution functions will differ from a Maxwellian distribution. When this effect becomes important, it is desirable to measure the temporal change of the velocity distribution functions of electrons and/or ions.

Fluctuations. Fluctuations and turbulence also have to be considered. In a gas, the only fluctuation that can occur, except for thermal fluctuations due to atomic and molecular motion, is fluid dynamic turbulence driven by the gas flow. A plasma not in thermal equilibrium, however, will move towards a lower energy state by a variety of energy relaxation processes. The free energy released during relaxation can drive various kinds of fluctuations. For example, plasma transport phenomena such as electrical resistance, thermal conductivity and diffusion often are determined by the nature of fluctuations that exist in the plasma. Fluctuations can occur in each of the quantities discussed above. The fluctuation in these values is usually written as the symbol for the parameter itself, with a $\sim$ attached, such as $\tilde{n}_e$, $\tilde{n}_i$, $\tilde{T}_e$, $\tilde{T}_i$, $\tilde{E}$ and $\tilde{B}$.

In the following sections, some characteristic plasma lengths and frequencies are listed using the basic parameters described above and simple explanations for each are given. In this explanation, only singly ionized ions are considered and A is the atomic mass number. The units for n_e, T_j and B are m^{-3}, eV, and T, respectively: all other quantities are in SI units.

Debye length. When a grid is inserted into a plasma, and a voltage applied to the grid, the ions and electrons will move so as to cancel the effect of the grid potential. Hence, the effect of the grid potential is felt only within a short distance from the grid itself. This distance is called the Debye length, or the Debye shielding distance.

The Debye length is given by

$$\lambda_D = \sqrt{\frac{\varepsilon_0 k T_e}{e^2 n_e}} = 7.43 \times 10^3 \sqrt{\frac{T_e}{n_e}} \quad (m) \tag{1.2}$$

and can be considered to depend essentially on the motion of the small, light electrons. If the electron temperature is zero, then λ_D is zero, and the potential is completely shielded, but if the electrons have finite thermal motion, the potential is shielded only partially. Similarly, if the electron density is infinitely large, λ_D is zero, but if n_e is finite, some 'leakage' of the potential occurs and λ_D is also finite. The size of the Debye length affects many types of plasma phenomena, including the wavelength dependence of the decay of plasma oscillations and the fundamental nature of Thomson scattering phenomena. Hence, it is extremely important in influencing both plasma phenomena and laser-aided plasma diagnostics.

Plasma frequency. In a plasma, the amounts of positive and negative charge are on average the same, and so the plasma is electrically neutral. If there is a small deviation from this equilibrium, however, the charged particles will move to neutralize the deviation, producing an oscillation around the average state called *plasma oscillation*. The characteristic frequency at which this oscillation occurs is called the *plasma frequency*, ω_{pj} and is given by

$$\omega_{pj} = \sqrt{\frac{n_j e^2}{m_j \varepsilon_0}} \tag{1.3}$$

where $j = e, i$. For laser measurements, the electron plasma frequency is particularly important, and this is given by

$$\omega_{pe} = 56.3 \sqrt{n_e} \quad (rad\ s^{-1}). \tag{1.4}$$

One of the most important points to remember about ω_{pe} is its relationship to the propagation of electromagnetic waves in the plasma. For conditions in which there is no magnetic field, or when the electric field of the wave is parallel to the magnetic field, electromagnetic waves with frequency less than ω_{pe} cannot propagate into the plasma. In these circumstances, the electric field of the waves is swamped by fluctuations of electrons in the plasma, and the situation is

Cyclotron frequency and Larmor radius. The motion of charged particles in a plasma can be found by solving the equation of motion (1.1). When a charged particle moves in a magnetic field, it is convenient to separate the motion into two components, one corresponding to circular motion around the magnetic field and one corresponding to the motion of the centre of the rotation. The rotational component is called *cyclotron motion*, due to its similarity to the motion of changed particles in a cyclotron accelerator, and the other component is called *drift motion*. The frequency and radius of the cyclotron motion are called the *cyclotron frequency* ω_c and the *Larmor radius* r_L respectively. The cyclotron motion of electrons and ions are quite different, due to the large difference in their masses. The values of ω_c and r_L are very different as is the direction of rotation. The relevant values of ω_c and r_L are given by

$$\omega_{cj} = \frac{eB}{m_j} \tag{1.5}$$

$$\omega_{ce} = 1.76 \times 10^{11} B \quad (\text{rad s}^{-1}) \tag{1.6}$$

$$\omega_{ci} = 0.96 \times 10^8 \frac{B}{A} \quad (\text{rad s}^{-1}) \tag{1.7}$$

$$r_{Lj} = \frac{v_j}{\omega_{ci}} \tag{1.8}$$

$$r_{Le} = 3.37 \times 10^{-6} \frac{\sqrt{T_e}}{B} \quad (\text{m}) \tag{1.9}$$

$$r_{Li} = 1.45 \times 10^{-4} \frac{\sqrt{AT_i}}{B} \quad (\text{m}). \tag{1.10}$$

Thermal velocity. Particles in a plasma that have temperature T move with an average speed, related to that temperature, called the thermal velocity. Even if electrons, ions, atoms and molecules have the same temperature, their thermal velocities are quite different because of the large differences in mass. The thermal velocities v_{th} are given by

$$v_{thj} = \sqrt{\frac{3kT_j}{m_j}} \tag{1.11}$$

$$v_{the} = 7.26 \times 10^5 \sqrt{T_e} \quad (\text{m s}^{-1}) \tag{1.12}$$

$$v_{thi} = 1.69 \times 10^4 \sqrt{T_i/A} \quad (\text{m s}^{-1}). \tag{1.13}$$

1.2.3 Summary

Based on these descriptions, the range of densities and temperatures in plasmas

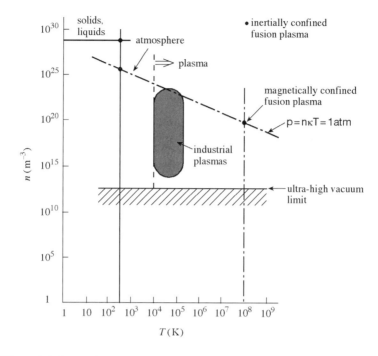

Figure 1.1. States of matter plotted in coordinates of temperature T and density n.

can be summarized in the manner shown in figure 1.1. When the temperature of the gas reaches above a few thousand Kelvin, the gas becomes ionized and a plasma is formed. Although there is a wide range of temperatures for plasma states, from the viewpoint of applications, the temperature T and the density n have upper and lower limits. Most types of plasmas used for industrial applications are operated either at or at less than atmospheric pressure. Because $p = nkT \leq 1$, this places an upper limit on the density n that can be obtained for a given temperature T. The desired conditions for plasmas used for magnetically confined fusion are $T \sim 10^8$ K and $n \sim 10^{20}$ m^{-3}. It is interesting to note that this type of plasma still lies on the $p \sim 1$ atm line shown in the figure, although this might be only a coincidence. The lower limit for electron density is determined by the pressure that can be achieved by the vacuum pump used in each application. The minimum density of certain species, determined by this condition, is of the order of 10^{12}–10^{13} m^{-3}.

It can be seen from figure 1.1 that the first three types of plasma listed in section 1.2.1 are plasmas with temperatures of around 10 000 K, while the fourth type, magnetically confined fusion plasma, has a much higher temperature of around 100 000 000 K. There are very few examples of industrially useful plasmas in the intermediate range of temperatures, and one consequence of this is that research concerning thermonuclear fusion has become quite isolated from the rest of the plasma research. Hence, it is difficult to directly relate the results of fusion research to other active plasma research areas, and vice versa. Recently, however, there have been a few exceptions to this. The electron cyclotron resonance heating (ECRH) mechanism, which was developed for fusion research, has been used to create a new type of high-density plasma source, called an ECR plasma, for use in semiconductor production. Also, the Thomson scattering measurement technique that was developed to measure electron properties in fusion plasmas has been applied recently to industrially useful plasmas.

In the course of the next 20 or 30 years, it is probable that new and active research fields will emerge in the region between current low-temperature plasma fields and thermonuclear fusion.

1.3 Different States of Matter and their Kinetic Properties

In the section above, we considered some important aspects of gases and plasmas. This section is a summary of the basic properties of all materials, including solids and liquids, with emphasis on the similarities and differences between the different states of matter.

Table 1.1 lists the various states of matter, and shows the material properties of each type of matter, together with the kinetic characteristics that describe its motion. With regard to material properties, the most important distinction between the types of matter is that solids and liquids can be considered to be 'condensed matter' while gases and plasmas are 'non-condensed matter'.

In condensed matter, the kinetic motion of individual atoms in the material is constrained by the potential energy that exists between neighbouring atoms. The distance between the atoms ($\sim 5 \times 10^{-10}$ m) is of the same order as the particle size, and the particle density is about 10^{28} m^{-3}. Hence, it is difficult to increase the density by raising the pressure. For non-condensed matter, however, the kinetic energy of atoms (and other particles, such as electrons and ions in plasmas) is much greater than the potential energy between the particles. In other words, the distance between neighbouring particles is far greater than the size of the particles themselves. Hence, for gases and plasmas, the particle

Table 1.1. Material and kinetic characteristics of the four states of matter.

	Solids	Liquids	Gases	Plasmas
Material properties	Condensed		Uncondensed	
Kinetic properties	Rigid body mechanics	Fluid mechanics		
				Influence of E and B

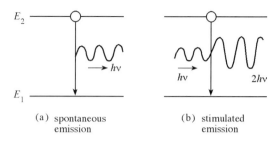

Figure 1.2. Spontaneous and stimulated emission for a two-level system.

density can be changed simply by altering the pressure. Other material properties, such as thermal and electrical conductivity, differ much both within the states and between the four states, and it is difficult to make general comments about these properties.

It is also possible to try and classify matter according to its kinetic characteristics. In these terms, solids and liquids have few common features. Rather, it is liquids, gases and plasmas that are similar. In most cases, fluid dynamics can be used to describe the kinetic motion of these three types of matter. The characteristic feature of each state of matter is incompressibility for liquids, compressibility for gases, and both compressibility and electrical conductivity for plasmas. The fact that plasmas have electrical conductivity is the reason that they can interact strongly with electromagnetic fields.

By looking at table 1.1, it can be seen that, with regard to material properties, gases and plasmas share the property of compressibility, and that with regard to kinetic properties, both can be described by fluid theory. Hence, it can be understood that gases and plasmas are similar in many important ways.

1.4 Characteristics of Laser Light

1.4.1 Coherence

The word *laser* comes from the initials of *l*ight *a*mplification by *s*timulated *e*mission of *r*adiation [3,4]. The light observed from atoms and molecules that have been excited to high energy levels, such as those in a discharge tube, is called spontaneous emission. This type of emission process is shown in figure 1.2(*a*). For spontaneous emission, a photon with frequency ν is emitted, where

$$h\nu = \Delta E \tag{1.14}$$

in which h is Planck's constant and ΔE is the energy difference between the upper and lower level of the transition. Both the phase and direction of spontaneously emitted photons are randomly distributed. This kind of light is called *incoherent*.

Another emission process, called *stimulated emission*, is shown in figure

1.2(b). In this process, a photon whose energy is in resonance with an energy transition in the atom or molecule triggers the emission process. The emitted photon has the same phase and the same propagation direction as the photon that triggers the emission. Light with these properties is called *coherent*. This stimulated emission process can be used to coherently amplify light. In a laser oscillator, an amplifier is placed in an optical resonator formed by two parallel mirrors, so that there is optical feedback to the amplifier. This is exactly the same principle that is used in electronic oscillators, and enables the laser oscillator to generate coherent light.

One of the most remarkable features of laser light sources is their high degree of coherence. This level of coherence was unattainable with previously available light sources. Oscillators using electronic devices such as vacuum tubes or transistors can generate coherent electromagnetic waves at radio and microwave frequencies, but there were no coherent sources at all in the visible part of the spectrum. Traditional light sources, such as incandescent lamps and discharge tubes, emit light that has noise-like fluctuations in phase and frequency. The coherence of the laser source is an important aspect of nearly all laser diagnostic methods and, for this reason, the concept of coherence is discussed in more detail below.

Temporal coherence. Electromagnetic waves are composed of mutually perpendicular electric and magnetic fields. In general, electromagnetic waves travelling in free space have only transverse field components, and the propagation of these waves can be described by a wave equation derived from Maxwell's electromagnetic theory. The most basic type of electromagnetic wave is a plane wave described by the equation

$$E(z,t) = E_0 \sin(\omega t - kz) \tag{1.15}$$

where E_0 is the electric field amplitude and ω is the angular frequency of the wave. The phase velocity c is related to the propagation constant k by

$$c = \frac{dz}{dt} = \frac{\omega}{k}. \tag{1.16}$$

It is possible to describe any arbitrary electromagnetic wave as a superposition of plane waves with different frequencies and propagation directions.

It should be noted that equation (1.15) is really valid only for electromagnetic waves that are perfectly coherent. Incoherent radiation consists of waves that fluctuate randomly in phase and frequency, and it is difficult to describe the behaviour of incoherent light with a simple wave equation. Historically, there was long debate about whether or not light was a wave phenomenon. One of the reasons for this long debate was that it is difficult to observe most common wave phenomena such as *diffraction* and *interference* with incoherent sources, and traditional light sources have very poor coherence. Wave phenomena, however, can be easily observed with coherent light sources.

The level of coherence for a particular wave is indicated by the degree to

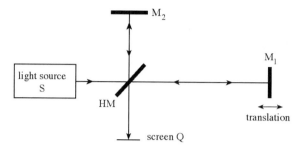

Figure 1.3. Schematic diagram of a Michelson interferometer.

which equation (1.15) describes the wave's motion. Because the frequency of light is very high, however, it is impossible to observe oscillation directly using an instrument such as an oscilloscope. The coherence of an electromagnetic wave can be estimated, however, using interference experiments such as those discussed below.

An *interferometer* is an instrument that can be described in general terms as a device in which light from a source is divided into two parts that are directed along different optical paths, and then recombined to form interference fringes. An example of such an instrument is the Michelson interferometer shown in figure 1.3. Light from a source is divided into two parts by the half-mirror, and interference of the light reflected from the mirrors M_1 and M_2 is observed on the screen. This situation can be considered mathematically as follows. In this discussion, the two optical path lengths are s_1 and s_2, and the electric field of the light wave has amplitude and frequency of E and ω, respectively. If the light is completely coherent, the intensity I at the screen is given by the product of the instantaneous amplitudes of the two waves. This can be expressed as

$$I \propto \langle E^2 \{\cos(\omega t - ks_1) + \cos(\omega t - ks_2)\}^2 \rangle = E^2 \{1 + \cos(s_2 - s_1)\}. \quad (1.17)$$

In this expression, the $\langle \rangle$ symbol represents time averaging. When the position of the mirror M_1 is changed, the optical path difference, $s_1 - s_2$, will change, and hence the intensity will vary between 0 and $2E^2$. This means that extremely high contrast interference fringes will be observed at the screen as the mirror M_1 is moved over a distance of the order of half a wavelength.

However, if the light has poor coherence, or the spectral width is broad, the contrast of the fringes will be diminished because the position of the fringes will be different for different frequency components of the light. Although interference will still be visible when the optical path length is zero ($s_1 = s_2$), the contrast will become poorer gradually as the distance $s_1 - s_2$ becomes larger. The optical path difference, $s_1 - s_2$, at which the contrast vanishes is called the *coherence length*. This length is an indication of the degree of coherence of the light source.

The coherence length for a light source with a frequency width $\Delta\omega$ is given by $2\pi c/\Delta\omega$. For example, the spectral width of the D lines from a sodium lamp is about 2 GHz, determined by Doppler broadening, and this corresponds to a coherence length of about 150 mm. Light from a luminescent solid, such as a radiating blackbody, has an extremely large spectral width and so its coherence length is correspondingly very short. By comparison, the coherence length of a frequency-stabilized laser can be as large as several hundred kilometres.

This discussion can be summarized fairly simply. A highly coherent laser produces electromagnetic waves that have a sinusoidal temporal oscillation. This means that the light has a very narrow spectral width and a very long coherence length. This type of coherence is called *temporal coherence*.

Spatial coherence. When a light source is spread over a broad spatial region, there is not necessarily any correlation in phase between light from different positions in the source, even if the light emitted from each point has good temporal coherence. When a light source is spatially extended, the degree of coherence can be estimated by placing a slit in front of the light source and observing interference fringes. This type of coherence is called spatial coherence.

For a wave propagating through space, a surface on which all the points have the same phase at a particular time is called a *wavefront*. A wave described by equation (1.15) has a wavefront in the *x–y* plane, and is called a *plane wave*. Light coming from a point source will have a spherically shaped wavefront, and this wave is called a *spherical wave*. A wave that possesses such a clear wavefront has a high degree of spatial coherence. In contrast, light from a source such as a discharge tube has poor spatial coherence, because the atoms or molecules in the discharge tube emit light with completely random phases, and so such a wavefront is not formed. Even in this case, it is possible to improve the spatial coherence by placing a small aperture in front of the discharge tube. However, this has the effect of reducing the effective area of the emitting region, and hence decreasing the light intensity. In the case of a laser, the light also is emitted from a medium with a finite cross-sectional area, but because the atoms or molecules emit light due to stimulated emission, the light from different parts of the laser medium is in phase and hence the spatial coherence is high.

A wave that has equal amplitude and equal phase at all points in the plane perpendicular to the propagation direction is called a perfect plane wave. This is the type of wave described by equation (1.15). In reality, however, such a perfect plane wave does not exist. Even light from a laser beam is finite in extent because the beam diameter is restricted by the diameter of the laser medium. This finite extent means that diffraction definitely will occur, and so transverse components of the wave will be generated. Hence, even a laser cannot generate a perfect plane wave.

The situation of a plane wave passing through an aperture is shown in figure 1.4(*a*). The plane wave becomes restricted in size by passing through the

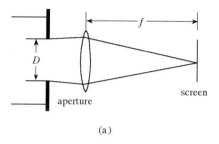

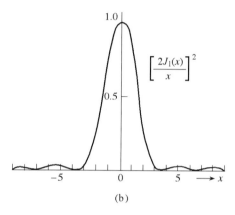

Figure 1.4. Fraunhofer diffraction due to a circular aperture.

aperture with diameter D, and the radial distribution at a distance sufficiently far from the aperture is given by *Fraunhofer's diffraction theory* [5] as

$$A(x) = [2J_1(x)/x]^2. \quad (1.18)$$

In this expression, $x = k\theta D/2$, J_1 is the first-order Bessel function and k is the wavenumber. The distribution $A(x)$ is shown in figure 1.4(b). Most of the energy of the wave is concentrated in a central disk. The first zero of $J_1(x)$ is at $x = 3.83$, and so the divergence angle due to diffraction by the aperture is $\theta \sim 1.22\lambda/D$. The distribution shown in figure 1.4(b) can be observed experimentally by focusing the beam with a lens and placing a screen at the focal plane of the lens. When this is done for a lens with focal length f, the relationship between the distance R from the centre of the beam and the beam angle θ is given by $R = f\theta$. Therefore, it is possible to focus a beam of diameter D down to a beam of size $R_0 \sim 1.22\lambda f/D$.

A well-formed wavefront with a high degree of spatial coherence, such as that produced by a laser, can be transformed into a plane wave or a spherical wave by using appropriate optics. In order to make a plane wave from a beam with $\lambda = 1$ μm and $D = 1$ mm, the beam divergence must be ~ 1 mrad, according to the above equation. In order to improve the directivity of a laser,

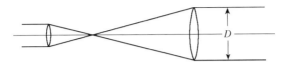

Figure 1.5. Schematic diagram of a beam collimator.

usually a beam collimator such as that shown in figure 1.5 is used to expand the beam diameter. For a beam with diameter of 1 m, a highly directional beam with divergence of about 1 μrad can be produced. It should be noted, however, that this kind of beam can be produced only if the laser is oscillating in a single transverse mode. This will be discussed in a later section.

A spherical wave can be produced by focusing a parallel beam with a convex lens. The smallest radius that can be produced, R_0, is limited by the spatial coherence of the light. For a beam with $\lambda = 1$ μm and a lens with $f/D = 1$, the minimum radius that can be produced is $R_0 = 1.22$ μm. The limit on focusing by a lens, corresponding to the smallest laser beam spot size that can be produced, is easy to remember because it is about the same order as the wavelength. This is called the *diffraction limit*.

1.4.2 Short Pulse Generation

The main advantages of lasers are listed in table 1.2. From this table, it can be seen that the most important keyword that separates lasers from other light sources is 'coherence'. When the word 'laser' is used, most people think of a strong, bright light source. In some applications, such as material processing, intense powerful lasers actually are used. However, even in this case, it should be noted that it is the high power *density* of the laser light that is important, and this comes from the laser's good focusing ability. For example, when a 1 W laser beam with wavelength of 1 μm is focused down to the diffraction limited size with $f/D = 1$ lens, the power density is extremely large: 2.1×10^{11} W/m². Many people would consider a 100 W light bulb to be a reasonably bright light source, but CO_2 lasers with power of only 20 W are used in laser surgery to cut

Table 1.2. Advantages of laser light.

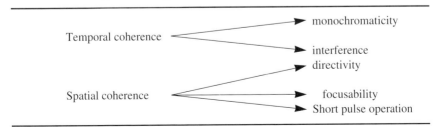

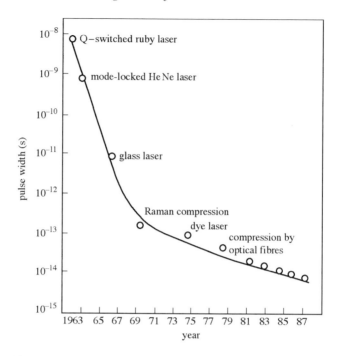

Figure 1.6. Chart showing the development of short laser pulses.

human skin. This is possible because of the high energy density that can be obtained by focusing the laser.

Although the upper limit of *average* power for commercially available lasers is about 30 kW, much higher *instantaneous* powers can be achieved from pulsed lasers. Lasers always have been at the forefront of ultrashort pulse technology, and figure 1.6 shows a history of short laser pulse generation. The current record for short pulse generation is 4 fs (4×10^{-15} s). As the period of one cycle of visible light is about 1 fs, the light wave in this short pulse oscillates only a few times during the pulse. In electronics technologies, the shortest pulses have lengths of the order of 10 ps, but much shorter pulses are achievable in laser technology. Generation of these ultrashort pulses is possible because the frequency of light is very high.

Commercially available Q-switched Nd:YAG lasers have typical pulse lengths of about 5 ns, and output powers of 1–2 J. This corresponds to instantaneous powers of 200–400 MW. Recently, it has become possible to generate terawatt order powers (1 TW = 10^{12} W) using table-top sized Ti:sapphire lasers that generate femtosecond pulses. By focusing the beams, extreme power densities of 10^{20} W m^{-2} order can be achieved. Power at this level causes all material to be instantaneously transformed into a high-density plasma.

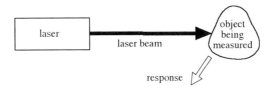

Figure 1.7. Schematic diagram of an active measurement method that uses a laser.

1.5 Advantages of Laser-Aided Measurement Methods

Lasers are used for a variety of purposes in a wide variety of fields that cover virtually all areas of science and technology. Areas in which laser-aided measurements are important are especially widespread, and include basic areas of physics and chemistry together with areas as diverse as electrical and mechanical engineering, process engineering, civil engineering, material science, geology, agriculture, medicine and biology. It is difficult to give a complete overview of all the characteristics of laser-aided diagnostic techniques, but before moving on to chapters dealing with specific research fields, it is useful to outline the basic characteristics that are usually thought of when laser-aided diagnostic methods are considered.

Ability to make 'active' measurements. In active measurement techniques, the object that is to be measured is stimulated in some way, and the response to that stimulation is observed. Laser-aided measurement methods are an example of such active techniques, and a basic experimental arrangement is shown in figure 1.7. The object to be measured is illuminated by the laser light and the response to the light is detected. The type of response differs according to the object being measured. The most typical response that is measured is a change in the laser light itself, in the form of a modulation of its amplitude, phase, frequency, polarization, or direction of propagation. Other responses include the generation of sound or heat, and the production of new particles, such as ions or new molecular species. In each case, the degree of the response is related to some property of the medium. Most of the advantages of laser-aided measurement methods, discussed below, are directly related to the fact that the techniques are active methods that use highly coherent laser beams.

High resolution and accuracy. Although laser measurement techniques do not *always* possess high resolution and high accuracy, most laser-aided methods have significantly better resolution and accuracy than passive optical methods. In fact, the difference in resolution and accuracy between active and passive methods usually is enormous, often as much as several orders of magnitude.

The main reason that laser-aided methods often have good resolution and high accuracy is based on the simple fact that lasers are highly coherent light sources. Because the light is coherent, the properties of the light, such as the amplitude, phase, polarization and direction of propagation, are constant, with only very small fluctuations. Hence, even very small changes in these quantities can be detected. In addition, because the laser energy can be concentrated into a small spatial region, a large signal from a small volume can be obtained, and this in turn leads to large signal-to-noise ratios.

Amongst all the possible responses, changes in phase can be detected most sensitively. As an example of this, consider the laser interferometer shown in figure 1.3. Distance can be measured with high accuracy by detecting the very small phase difference between the light that returns from mirrors M_1 and M_2. Interferometry is the most accurate method of measuring distance, and by using a heterodyne system in which the beams have different frequency, phase changes as small as 1/100 of a wavelength can be detected. This means that an interferometer can measure a distance of about 1 m with the accuracy of about 1 nm (10^{-9} m). Long optical path length interferometry has been used in an attempt to detect gravity waves, but in this case, the required accuracy is said to be of the order of 10^{-18}–10^{-20}. This kind of high-accuracy interferometry is an example of a measurement technique that exploits the coherence properties of lasers to their limit. In addition to distance, interferometry can be used to determine quantities such as changes in temperature and density in gases, distortion of solids, and electron density in plasmas. These quantities are determined by measuring the refractive index of an object placed inside one of the arms of the interferometer.

By changing the laser in figure 1.7 to a tunable laser, and measuring the signal as a function of excitation wavelength, the *excitation spectrum* can be obtained. The broad range of techniques in which such a spectrum is measured are collectively called *laser spectroscopy*. These techniques are probably the most important techniques that are described in this book. Compared with conventional spectroscopic techniques, in which the resolution of the measurement is determined by the resolution of the spectrometer used for the measurement, the resolution of laser spectroscopic techniques is determined by the spectral properties of the laser. A frequency-stabilized laser can generate monochromatic light that has spectral width of less than 500 kHz. However, it is rarely possible for this extremely narrow spectral width to be exploited fully. The spectral widths of most phenomena that are measured in gases and plasmas usually are determined by Doppler broadening caused by particle motion, and are of the order of 1 GHz in low pressure gases. Therefore, extremely high spectral resolution in laser spectroscopy is possible only using the technique of *Doppler-free spectroscopy*.

In many cases when laser spectroscopy is used for particle detection, high spatial and temporal resolution can be achieved simultaneously. The high focusability and directivity of a laser can be used to create a high degree of excitation within a small volume. This enables large signal-to-noise ratios to be obtained, and allows sensitive, high resolution measurements to be performed.

High degree of spatial and temporal resolution. In laser-aided measurements, extremely high values of spatial and temporal resolution can be achieved because the resolution is determined by the properties of the laser beam. High temporal resolution can be achieved by using a short laser pulse so that the pulse length determines the temporal response of the experiment. Resolution down to femtosecond order is possible. High spatial resolution can be achieved by focusing the laser beam so that the detected signal comes from a very small region. The diffraction effects discussed earlier limit the ultimate size of the focused beam, but as this limit is the of order of the laser wavelength, spatial resolution of micrometre order is possible.

Although temporal and spatial resolution of femtosecond (10^{-15} s) and micrometre order are achievable, the amount of detected signal becomes extremely small under these circumstances. Hence, some type of spatial or temporal averaging usually is performed in order to achieve good signal-to-noise levels. For measurement of the temporal change in a signal, a repetitively pulsed laser most often is used. It also is possible to use a cw laser and detect the change in the continuously detected signal, but in this case the temporal response of the detection system determines the temporal resolution of the experiment. For measurement of a spatial distribution, it is necessary to translate either the laser beam or the object, as is shown in figure 1.8(*a*). This technique is called *laser microscopy*.

Ability to make spatial imaging measurements. One of the key advantages of optical techniques compared with other methods is that spatial imaging measurements are possible. The most common examples of optical imaging

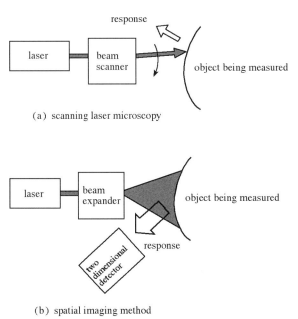

Figure 1.8. Two different methods of obtaining two-dimensional information.

devices are the camera and the human eye. In these cases, information is recorded simultaneously on all parts of a surface, such as the film. When a laser is used to obtain two-dimensional information, a scanning system such as that shown in figure 1.8(*a*) often is used, but this is not true spatial imaging. The system shown in figure 1.8(*b*) is such a system, in that the laser illuminates the entire area being observed and the signal from all parts of the illuminated surface is recorded in parallel. Note that in this case, the temporal and spatial resolution of the apparatus is determined by the detection system. Other instruments, such as the interferometer shown in figure 1.3, also can be used for imaging purposes. Holography is a good example of this kind of an imaging technique.

When a laser is used in an imaging experiment, the performance of the detector becomes very important, because the spatial resolution of the experiment is determined by the detection system. Photographic film has good spatial resolution and high sensitivity, but is not used commonly now because data analysis cannot be performed in real time. The development of high quality diode arrays and charge-coupled device (CCD) cameras has overcome this limitation, and such detectors are used routinely now for imaging purposes. When image intensifiers are combined with these detectors, sensitivity down to photon counting levels can be achieved.

Ability to make non-perturbing measurements. Another advantage of optical techniques in general is their non-intrusive, non-perturbing nature. In addition, the use of lasers enables the light source to be located even further from the object to be measured. Although a window to allow the laser light to enter the chamber and a window to enable observation of the signal are necessary, in-situ non-perturbing measurements are possible. This is one of the principal advantages of laser measurement techniques. When extremely intense laser light is used, however, it is usually necessary to ensure that the light does not perturb the conditions of the medium. Another advantage of optical techniques is that the systems usually are unaffected by electromagnetic interference.

Nonlinear optical techniques. Nonlinear optical effects are phenomena in which an intense electric field, due to intense light, induces a nonlinear response in a material. These effects were discovered by the use of intense laser light. From the viewpoint of classical electromagnetism, these effects are caused by the nonlinear response of the electrical susceptibility (polarization) of a medium, induced by the intense electric field. From the viewpoint of quantum mechanics, these phenomena can be explained in terms of multi-photon processes. These processes are used practically for the generation of higher harmonics and the frequency conversion of laser light. The concept of optical nonlinearity is central to both laser research and laser applications, and 'nonlinearity' can be added to 'coherence' as one of the most important keywords in laser engineering. The laser oscillator itself can be considered to be a kind of nonlinear device.

Remote sensing capability. Active remote sensing over a range of a few 100 km is possible by transmitting a laser beam into the atmosphere, and measuring the response. In this case, the signal is due to light that is backscattered from particles in the atmosphere and information about the distribution of the particles can be obtained. If, instead of measuring backscattered light, a reflector is set in a remote place, the receiving signal intensity is increased and the distance to the mirror can be determined. The longest distance measured in this way is the distance from the Earth to the moon.

Conventional radar (*r*adio *d*etection *a*nd *r*anging) uses microwave sources, and the spatial distribution of an object can be determined by time-analysing the echo received when a pulsed electromagnetic wave is transmitted. When the microwave source is replaced with a light source, the technique is called *laser radar*, or *LIDAR* (*l*ight *d*etection *a*nd *r*anging). The advantages of using light as the electromagnetic radiation are that highly sensitive detection of atmospheric particles is possible by Mie and Rayleigh scattering, and spectroscopic methods can be used to measure the concentrations of different species of particles. For these reasons, LIDAR has become a very useful technique for environmental and atmospheric monitoring. It also is used in fusion research. In this field, picosecond laser pulses are used to detect Thomson scattering with a spatial resolution of less than 100 mm.

As well as the types of remote sensing mentioned above, the type of optical fibre used for optical communication can be used for remote-sensing measurements. One example of this is fibre Doppler flowmetry, in which the flow of blood in a human vein is measured. This is an example of fibres being used to access regions that are difficult to probe by any other means. The resistance of fibres to electromagnetic effects also allows them to be used as sensors of electric fields and magnetic fields in high-voltage environments. These devices are called optical potential transformers (PT) and optical current transformers (CT).

REFERENCES

[1.1] Spitzer L Jr 1962 *Physics of Fully Ionized Gases* (Interscience)
[1.2] Chen F 1984 *Introduction to Plasma Physics and Controlled Fusion* vol 1 (Plenum)
[1.3] Arecchi F T *et al* ed 1972–85 *Laser Handbook* vol 1–5 (North-Holland)
[1.4] Maeda M 1987 *Quantum Electronics* (Shokodo) (in Japanese)
[1.5] Born M and Wolf E 1974 *Principle of Optics* (Pergaman)

Chapter 2

Basic Principles of Different Laser-Aided Measurement Techniques

When electromagnetic waves enter a gas or plasma, a number of different phenomena can occur, and these can be classified in the four ways shown in figure 2.1: reflection, transmission, refraction and scattering. The degree to which each phenomenon occurs depends on the parameters that characterize the gas or plasma together with the properties of the incident laser light, such as its wavelength, spectral structure, pulse length and intensity. In laser-aided measurements, the changes in the laser radiation due to the gas or plasma are measured quantitatively. Information about the gas or plasma then is obtained using known relationships between the induced changes in the laser radiation and the plasma parameters. In order to make high-quality measurements, it is necessary to use a laser that has properties that produce effects that are large enough to measure. The quality of any particular measurement usually is determined by the suitability of the laser source.

The phenomena shown in figure 2.1 can occur for radiation with wavelengths ranging from nanometre to millimetre order, and the appropriateness

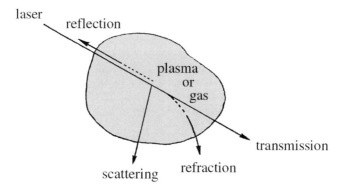

Figure 2.1. Phenomena that occur when a laser propagates through a gas or plasma.

Interaction of Electromagnetic Waves with Single Particles 27

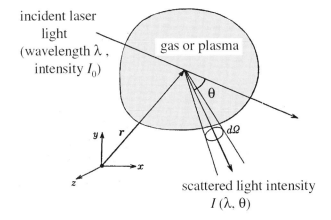

Figure 2.2. Scattered light from a laser passing through a gas or plasma.

of a laser source for a particular measurement depends on the properties of the gas or plasma being measured. Microwave wavelength sources have been available for many years, but most of the recent progress in laser diagnostic methods is due to progress in laser sources, especially those at shorter wavelengths.

The first section of this chapter is an introduction to the general effects of electromagnetic radiation on single particles in gases and plasmas. This is the basis of all the effects shown in figure 2.1. Phenomena such as Thomson scattering, Rayleigh scattering, Raman scattering, resonant absorption and fluorescence are described briefly. Section 2.2 is a description of effects due to collections of particles (i.e. the entire gas or plasma), and contains explanations of reflection, transmission, refraction and scattering. In section 2.3, detailed descriptions of spectral profile measurements are given.

2.1 Interaction of Electromagnetic Waves with Single Particles

When an electromagnetic wave interacts with a particle, as shown in figure 2.2, a differential scattering cross-section can be defined as follows. For laser radiation with wavelength λ_0 and intensity I_0 (W m^{-2}), the amount of scattering that occurs at wavelengths between λ and $\lambda + \Delta\lambda$, at an angle θ, into a solid angle $\Delta\Omega$, has intensity $I(\lambda,\theta)$ that can be written as

$$I(\lambda,\theta)\mathrm{d}\Omega\mathrm{d}\lambda \propto I_0 n V \mathrm{d}\Omega\mathrm{d}\lambda \tag{2.1}$$

where n is the density of the scattering species and V is the scattering volume. The proportionality constant in this equation is called the *differential scattering cross-section*, and is given the symbol $\sigma(\lambda,\theta)$. Using this, equation (2.1) then can be rewritten as

$$I(\lambda,\theta)\mathrm{d}\Omega\mathrm{d}\lambda = I_0 n \sigma(\lambda,\theta) V \mathrm{d}\Omega\mathrm{d}\lambda \tag{2.2}$$

28 Basic Principles

The size of the scattering cross-section $\sigma(\lambda,\theta)$ depends on the laser wavelength, the type of scattering particle and the properties of the plasma or gas. There are several different types of scattering, mainly differentiated by the type of scattering particles. These are discussed separately below.

2.1.1 Thomson Scattering by Charged Particles

The acceleration of a charged particle of mass m and charge q by a plane electromagnetic wave with electric field $E(r,t)$ is given by

$$\dot{v} = \frac{q}{m_j} E(r, t). \tag{2.3}$$

A charged particle that is accelerated will radiate energy, and for the case considered here, when the particle velocity is much smaller than the speed of light, the radiated energy per unit time I_a^S is given by

$$I_a^S = \frac{q^2}{6\pi\varepsilon_0 c^3} \dot{v}^2. \tag{2.4}$$

This type of radiation from a charged particle is called *Thomson scattering* [1]. One important point to note is that $|\dot{v}_e| \gg |\dot{v}_i|$, because $m_e \ll m_i$ and so the amount of Thomson scattering from an ion is much smaller than that from an electron. Hence, when considering scattering from a plasma, which contains both electrons and ions, it is necessary to consider only the scattering from electrons.

The electron acceleration occurs in the direction of the electric field of the incident radiation, and so the scattered light has a characteristic direction of polarization. The relationship between the incident and scattered waves is

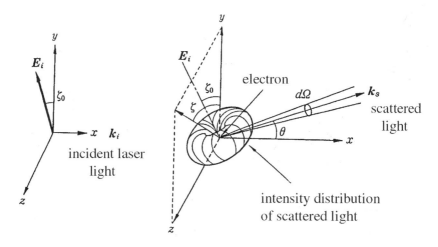

Figure 2.3. The directional distribution of the Thomson-scattered light intensity for the case of linearly polarized incident light with electric field E_i.

shown in figure 2.3 for the case of linearly polarized incident light. The angle between the incident wavevector k_i and the detected scattered wavevector k_s is called the *scattering angle* and denoted by θ. The distribution of scattered radiation has an apple-shape, as shown in the figure. One point to note is that there is no radiation in the direction of E_i.

The differential cross-section for Thomson scattering $\sigma(\lambda_i, \theta)$ is independent of the wavelength of the incident light λ_i. The value of $\sigma(\lambda_i, \theta)$ for incident radiation with polarization direction ζ is given by

$$\sigma(\theta) = r_0^2 \left[1 - \sin^2\theta \cdot \cos^2(\zeta_0 - \zeta)\right]. \qquad (2.5)$$

In this expression, r_0 is the *classical electron radius*, given by

$$r_0 = \frac{e^2}{4\pi\varepsilon_0 m_e c^2} = 2.82 \times 10^{-15} \text{ m}. \qquad (2.6)$$

For the case when the incident radiation is unpolarized, the cross-section can be obtained by integrating equation (2.5) over ζ, and becomes,

$$\sigma(\theta) = \frac{r_0^2 (1 + \cos^2\theta)}{2}. \qquad (2.7)$$

The total Thomson scattering cross-section σ_T is given by

$$\sigma_T = \int \sigma(\theta) d\Omega = \frac{8}{3}\pi r_0^2 = 6.65 \times 10^{-29} \text{ m}^2. \qquad (2.8)$$

2.1.2 Mie and Rayleigh Scattering

Mie scattering is scattering of electromagnetic radiation from a particle whose size, a, is of the same order as the wavelength of the radiation, λ. It is an elastic scattering phenomenon and the incident light and the scattered light have the same wavelength. The theory of Mie scattering is based on the diffraction of a plane electromagnetic wave from a small sphere with a complex refractive index, and can be derived from Maxwell's equations of electromagnetic theory [2]. This theory is based on the assumptions that the particle is a perfect sphere, the refractive index of the particle is known, and multiple scattering does not occur.

Figure 2.4 shows the angular distribution of Mie scattering. The parameter $\alpha = 2\pi a/\lambda$ is called the size parameter. It can be seen from the figure that when the particle is small, scattering in the forward and backward directions have about the same magnitude, but forward scattering becomes dominant as the particle size increases. Figure 2.5 shows an example of an intensity distribution measured for the case of scattering at 90° with $\lambda = 514$ nm. The two curves in the graph represent scattering from particles with refractive index of 1.69 (solid line) and $1.69 - 0.4i$ (dashed line). It can be seen that the overall trend is for the scattering to become smaller as the particle size decreases, but there are variations in this dependence. For $\alpha < 1$, the decrease is exceptionally fast, and the signal size depends on the sixth power of the particle size. For $\alpha > 2$, there

30   Basic Principles

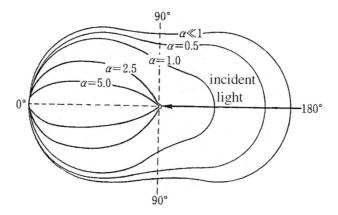

Figure 2.4. Angular distribution of Mie scattered light [5]. α is the scattering parameter and for this case, $m = 1.33$.

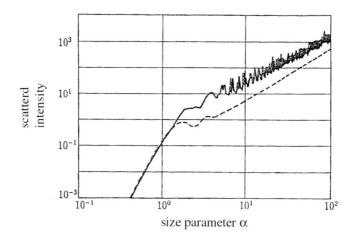

Figure 2.5. Example of results of a Mie scattering measurement. For this case, the laser wavelength was 514 nm and the scattering angle was 90°. The refractive indices are 1.69 (solid line) and $1.69 - 0.4i$ (dashed line).

is some fluctuation due to diffraction, but if this is ignored, the scattered signal depends on the square of the particle size. The dependence of the scattering intensity on radiation wavelength is slightly more complicated. For sufficiently small particles, the wavelength dependence is to the inverse fourth power, the same as that for Rayleigh scattering, discussed below. For larger particles, the dependence is much more complicated.

When the wavelength of the incident light is sufficiently large compared with the particle size, the dominant form of scattering is from the electric dipole induced by the incident light. This type of scattering is called *Rayleigh scattering*, and can be observed from gas molecules or atoms. Randomly distributed particles in a dielectric material also exhibit Rayleigh scattering.

Because the amplitude of the electric dipole moment is proportional to the product of the polarizability of the particle α and the incident light intensity I, the scattered light intensity I_r is given by the following formula,

$$I_r = \frac{8\pi^4 \alpha (1 + \cos^2 \theta)}{\lambda^4 r^2} I \tag{2.9}$$

where θ is the scattering angle, and r is the distance to the observation point. As can be seen from this equation, the Rayleigh scattering intensity is proportional to λ^{-4}, and the angular distribution is similar to that shown in figure 2.3.

2.1.3 Raman Scattering

Raman scattering is a form of inelastic scattering, and the wavelength of Raman scattering is different from that of the incident light. Scattering at longer wavelengths is called Stokes Raman scattering and that at shorter wavelengths is called anti-Stokes Raman scattering, as shown in figure 2.6. For the case of Raman scattering by a gas molecule, the Raman shift frequency is determined by the vibrational or rotational energy levels of the molecule. This means that identification of the scattering molecule is possible by measuring the Raman shift frequency. This technique is called *Raman spectroscopy*. For the case of crystals, the Raman shift frequency is determined by lattice vibrations.

The electric polarization P induced in a molecule with a vibrational coordinate q by an electric field E is given by

$$P = \varepsilon_0 \alpha_0 E + \varepsilon_0 \left(\frac{\partial \alpha}{\partial q}\right)_0 qE. \tag{2.10}$$

In this expression, α is the polarizability of the molecule. The Raman scattering intensity is proportional to the second term in this equation, and is usually much smaller than the Rayleigh scattering intensity. Table 2.1 shows Raman shift frequencies for various materials.

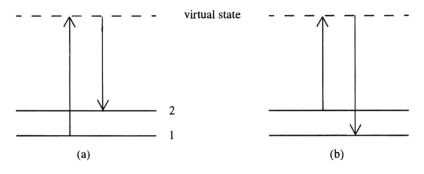

Figure 2.6. Energy level diagram showing the Raman scattering process for (*a*) Stokes Raman scattering and (*b*) anti-Stokes Raman scattering.

32   Basic Principles

Table 2.1. Examples of Raman shift frequencies for different materials.

Material	Phase	Raman shift (cm^{-1})
H_2	gas	4155
D_2	gas	2991
CH_4	gas	2916
O_2	gas	1552
Water	liquid	3420
Nitrobenzene	liquid	1345
Diamond	solid	1332
Quartz	solid	467

2.1.4 Resonant Absorption

When the frequency of light passing through a material exactly matches one of the resonant frequencies of a species in the material, the light will be absorbed. This process is called *resonant absorption*. In atoms, the resonant frequencies are due to electronic transitions. In molecules, the electronic transitions have complicated vibrational and rotational transitions superposed on them.

The intensity of absorption is given by the absorbance α or the absorption cross-section σ. According to the quantum mechanical theory, the absorption cross-section σ is related to the Einstein B coefficient by the following expression:

$$\sigma(\nu) = \frac{h\nu}{c} Bg(\nu) \qquad (2.11)$$

where ν is the frequency of light, $g(\nu)$ is the normalized line shape function and c is the speed of light. The Einstein B coefficient, in turn, is given by

$$B = \frac{|\mu_{12}|^2}{6\varepsilon_0 \hbar} = \frac{\langle u_1|e\mu|u_2\rangle^2}{6\varepsilon_0 \hbar} \qquad (2.12)$$

where μ_{12} is the matrix element of dipole moment $e\mu$, u_1 and u_2 are the wave functions of lower and upper levels, and ε_0 is the vacuum permittivity. Therefore, the absorption cross-section depends on μ_{12} and the line shape $g(\nu)$.

After an atom or a molecule has been excited to an upper energy level by absorption, relaxation to a thermal equilibrium state occurs, which releases the absorbed energy. Radiative relaxation is called *spontaneous emission* or *fluorescence*. The fluorescence intensity is one indication of the amount of absorption that occurs. The measurement technique based on this is called *fluorescence spectroscopy*. In cases when the relaxation is non-radiative, the local temperature rises, and this temperature change can be observed via the deflection of a light beam or generation of acoustic waves. Techniques based on these two phenomena are called *thermal lens spectroscopy*, and *photo-acoustic spectroscopy*, respectively.

The development of tunable lasers in the late 1960s revolutionized diagnostic techniques based on light absorption. These measurement methods, including absorption spectroscopy in which the absorption is measured directly, are discussed in section 2.2.

2.1.5 Photo-Ionization

Atoms and molecules can be ionized by exposing them to short wavelength light. This process in called photo-ionization. Table 2.2 shows the ionization energy of various elements. The ionization energies are higher than 4 eV and so ultraviolet (UV) light is required for photo-ionization to occur. If the incident radiation is x-ray or UV light, multiple electron ionization or detachment of inner-shell electrons can occur. If the incident light is very intense, nonlinear multi-photon ionization may occur. In this case, photo-ionization is possible even with visible or infrared (IR) radiation.

2.2 Laser Propagation through Gases and Plasmas

There are potentially many different species in gases and plasmas, including different types of charged particles, atoms, molecules and sometimes small particles. The propagation of laser light through a gas or plasma is affected by a combination of the elementary processes outlined above. These phenomena are discussed separately below.

2.2.1 Reflection

When a laser is introduced into a plasma or gas, any phenomena that results in radiation propagating in the reverse direction can be considered a form of reflection. However, it is possible for some fraction of the radiation to be scattered from particles in the gas or plasma in the reverse direction. This will be discussed in section 2.2.4, which concerns scattering processes. In this section, we will restrict ourselves to a narrower definition of reflection, in which the radiation is totally reflected in a mirror-like way from the plasma or gas. With this definition, reflection cannot occur for a gas but can for a plasma if the

Table 2.2. Ionization energies of atoms.

Atom	Ionization energy	Atom	Ionization energy
H	13.60 eV	Mg	7.64 eV
He	24.59	Al	5.99
Li	5.39	Si	8.15
C	11.27	Ar	15.76
N	14.53	K	4.34
O	13.62	Ca	6.11
Na	5.14	Fe	7.87

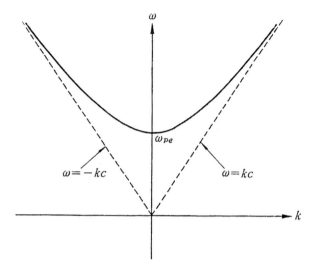

Figure 2.7. Dispersion relation for electromagnetic waves in a plasma propagating in the O-mode.

condition called *cut-off* is satisfied. This condition is defined as the refractive index of the plasma n ($= ck/\omega$) being zero.

In general, the propagation characteristics of waves in a medium are given by an expression called the *dispersion relation*. The dispersion relation in a plasma depends on many factors, including the wave frequency, the plasma density and temperature, and magnetic fields in the plasma. One way of simplifying this complex situation is to consider the plasma temperature to be zero and plot the normalized wave frequencies on a graph with density on the horizontal axis and magnetic field strength on the vertical axis. This is called an Allis diagram [3].

From the Allis diagram, the important characteristic wave frequencies, such as the plasma frequency, the cyclotron frequency, and regions where the dispersion relation changes rapidly, can be seen at a glance. The dispersion relation usually is shown graphically using k and ω as coordinates, in the way shown in figure 2.7.

The cut-off condition for reflection of waves from a plasma ($ck/\omega = 0$) is indicated in the diagram by a point where the tangent to the dispersion relation is horizontal. In a plasma, there are several different conditions for which the cut-off condition can be satisfied. Amongst these, we will consider only one important case, the cut-off for the O-mode, which is the wave mode in which the electric field of the wave is parallel to the magnetic field. In this case, the dispersion relation is given by

$$\omega^2 = \omega_{pe}^2 + k^2 c^2 \tag{2.13}$$

where $\omega_{pe} \equiv \sqrt{(n_e e^2 / m_e \varepsilon_0)}$ is the electron plasma frequency. Figure 2.7 shows

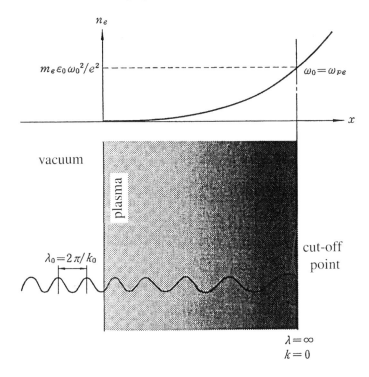

Figure 2.8. Electromagnetic waves with wavelength λ_0 ($=2\pi/k_0$) entering a plasma. The waves will be reflected at the cut-off point where $\omega_0 = \omega_{pe}$.

equation (2.13) plotted on a graph with k and ω as the coordinates.

In regions where $\omega > \omega_{pe}$, $k^2 > 0$ and so electromagnetic waves can propagate through the plasma. In particular, for the case of $\omega \gg \omega_{pe}$, the electromagnetic wave has a frequency much larger than the characteristic frequency of the electrons, ω_{pe}, and so the electrons are not affected by the wave electric field. When $\omega < \omega_{pe}$, however, $k^2 < 0$, and so the wave cannot propagate. In this case, the electrons can respond strongly to the wave electric field and this results in a flow of electron current at the surfaces around the plasma, in the same way that current flows on the surface of a conductor.

Now consider the case when the electron density has a gradient, and the electromagnetic wave enters the plasma from a region of low density. In a vacuum, the wave has wavelength λ_0, frequency ω_0 and wavevector k_0, where $k_0/\omega_0 = c$ and $\lambda_0 = 2\pi/\omega_0$. As the wave propagates into a region of increasing electron density, the electron plasma frequency ω_{pe} also increases. As shown in figure 2.8, the frequency of the wave ω will decrease until it reaches the point where $\omega = \omega_{pe}$. At this position, $k = 0$ and the wave cannot propagate any further. This is called the cut-off position. This effect has been used for many years in upper atmosphere research as a method to find the position of the reflection layer in the ionosphere. In this case, electron density is around 10^8–10^{12} m^{-3} and the cut-off frequencies are $f_{pe} = \omega_{pe}/2\pi = 0.1$–10 MHz. Hence, radio waves are used as the probing electromagnetic waves. In laboratory plasma research, this effect has been developed as a simple method of estimating the electron density in an arc plasma.

In recent years, this reflection effect has been combined with transmission interferometry, described in the next section, into a single instrument called a reflecto-interferometer, or more simply, a *reflectometer*. This instrument is used to measure electron density, and has become an important diagnostic tool in the high-density plasmas used in magnetically confined fusion plasma research. In the large-scale devices used in fusion research, plasmas are surrounded by neutron shields and blanket layers, the latter being used to convert neutron energy into thermal energy. The presence of these layers severely restricts the size and the number of diagnostic ports. A reflectometer only needs a small window port on the vacuum chamber, and so is particularly well suited for this environment.

2.2.2 Transmission

An electromagnetic wave that is transmitted through a plasma or gas can have its intensity I, phase ϕ and polarization ξ altered by the plasma or gas. The size and type of the changes, ΔI, $\Delta \phi$ and $\Delta \xi$, are determined by the properties of the medium as well as those of the electromagnetic wave. If the changes are large enough to be detected, then transmission methods that detect these changes can be used to measure the properties of the plasma or gas.

Measurement of ΔI. The intensity change ΔI of an electromagnetic wave due to transmission over a distance l through a medium containing absorbing particles with density n and absorption cross-section σ_{ab} is given by

$$\Delta I = I_0 - I_1 = I_0[1 - \exp(-n\sigma_{ab}l)] \qquad (2.14)$$

where I_0 and I_1 are the initial and final wave intensities, respectively.

The intensity of an electromagnetic wave passing through a medium is reduced due to each of the scattering processes described in section 2.1, and a cross-section σ_{ab} can be defined for each process. However, in nearly all types of gases and plasmas, scattering due to processes other than resonant absorption is very small. By matching the wavelength of the incident radiation to a resonant wavelength of particles in the gas or plasma, measurement of the amount of absorption ΔI can be used to determine the density of atoms, molecules or ions in the plasma.

For this to be valid, however, the spectral width of the laser must be narrower than that of the absorption line. For the case when $n\sigma_{ab}l \ll 1$, equation (2.14) simplifies to $\Delta I \sim I_0 n \sigma_{ab} l$. Hence, if l and σ_{ab} are known, then the density of absorbing particles n can be found by measuring ΔI. Progress in the development of tunable lasers capable of producing radiation from ultraviolet to

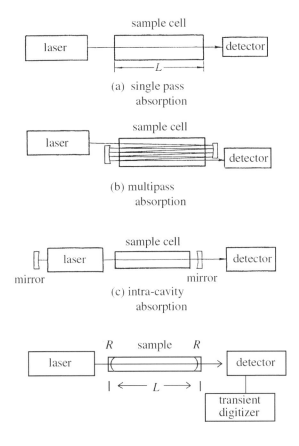

Figure 2.9. Different arrangements for absorption spectroscopy measurements.

infrared wavelengths has made this kind of measurement possible. If the spectral profiles of atoms and molecules are measured, other properties in addition to density can be obtained. Spectral profile measurements can be performed using both absorption and scattering methods, and are discussed in section 2.3.

The simplest arrangement for absorption spectroscopy is shown in figure 2.9(a). Absorption cross-sections for atomic gases are of the order of 10^{-17} m^2 and so, for a cell of length 1 m, a density of 10^{14} atoms/m^3 is needed to produce a reduction in laser intensity of 1%. This order of density can be considered as the detection limit for absorption measurements in atomic gases. For molecular gases, the lower detection limit is one or two orders of magnitude worse.

The simplicity of absorption methods makes them the standard methods for detection of high-concentration species. Figures 2.9(b) and (c) shows examples of absorption apparatus designed to produce higher sensitivity. Figure 2.9(b) shows a type of multipass cell called a *white cell*. It is possible to achieve more than 50 passes of the beam through this type of cell, and the consequent increase in absorption length L increases the sensitivity of the absorption measurement. Although multipass cells are effective at increasing the detection sensitivity, even better sensitivity can be obtained by placing the absorption cell inside the laser resonator itself. This is called *intra-cavity spectroscopy* and this type of arrangement can increase the sensitivity by more than three orders of magnitude. However, there are various experimental problems, such as alignment of the laser resonator, which limit the applicability of this method.

A relatively new method that has significantly better sensitivity than previous techniques is called *cavity ring-down spectroscopy* (CRDS) [4]. A typical experimental arrangement is shown in figure 2.9(*d*). In this technique, a short laser pulse is coupled into a high-Q optical cavity. The light intensity at the detector shows an exponential decay with decay constant τ given by

$$\tau = \frac{L}{c(\ln R + \sigma NL)} \qquad (2.15)$$

where L is the cavity length, R is the reflectivity of mirrors and c is the velocity of light. If the ring-down time τ can be determined with an accuracy of 10^{-3} and the reflectivity of the mirrors R is very close to 1 (i.e. $1 - R \leq 10^{-4}$), absorption coefficients of less than 10^{-7} m^{-1} can be determined using a 1 m cavity.

An extremely wide range of molecules and radicals can be detected by measuring absorption due to vibrational and rotational transitions. These transitions usually have wavelengths of more than a few micrometres. Until recently, the applicability of absorption methods in this wavelength range was limited because the only practical laser sources were Pb-compound laser diodes. However, the development of optical parametric oscillators and difference frequency mixers means that powerful infrared laser sources, tunable over a wide wavelength range, are now available, and IR absorption measurements are becoming widely used.

In addition to the resonant absorption by particles, another resonance condition in plasmas occurs when $n = ck/\omega = \infty$. When the dispersion relation is plotted in (k,ω) coordinates, in the way shown in figure 2.7, this resonance is indicated by the point where the tangent to the dispersion relation is vertical. Large absorption occurs when this condition $ck/\omega = \infty$ is close to being satisfied, and this effect is used as a basic mechanism to heat plasmas. For example, the electron cyclotron resonance (ECR) can be used as a plasma-heating method in both high-temperature fusion plasmas and low-temperature processing plasmas. Recently, there have been some attempts to measure electron density and temperature using this effect.

Measurement of $\Delta\phi$. The speed of electromagnetic waves in a gas or plasma is different to their speed in a vacuum. The difference is determined by the density of the gas or plasma, and is used as the basis of a measurement method called interferometry. In this method, an electromagnetic wave that has passed through

Laser Propagation through Gases and Plasmas 39

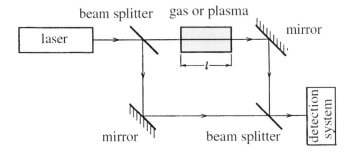

Figure 2.10. Schematic diagram of a Mach–Zehnder interferometer.

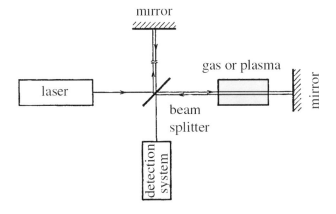

Figure 2.11. Schematic diagram of a Michelson interferometer.

a gas or a plasma is combined with a wave that has not. The phase difference between the two waves means that interference can be observed. Because phase differences of the order of 10^{-3} rad can be measured, interferometry is an extremely accurate and reliable method of measuring atomic, molecular and electron densities.

The types of instruments used for interferometer measurements can be divided into two main types: the Mach–Zehnder interferometer, shown in figure 2.10 and the Michelson interferometer, shown in figure 2.11. In both types, the laser beam is divided into two components using a mirror with 50% reflectivity. After one beam (called the measurement beam) has passed though the gas or plasma being studied, and the other beam (called the reference beam) has been transmitted through a vacuum or atmosphere, the two beams are recombined. The interference fringes produced by the two beams are detected by a photographic or photo-electric method, and shifts in the interference fringes produced by the plasma are analysed in order to obtain information about the plasma. In an actual interferometry experiment, the change in the fringes is measured, and so it is not necessary for the measurement beam and the reference beam to have exactly the same length, nor for the reference beam to be passed through the vacuum.

In the past, interferometry was performed using white light sources, and filters were used to produce radiation at the desired wavelength. However, the development of lasers has provided sources with intense coherent beams, with wavelengths ranging from the ultraviolet region to the microwave region. Most modern interferometers use a laser as the light source, and hence are called *laser interferometers*.

The principle of interferometry can be summarized as follows. When an electromagnetic wave of wavelength λ passes through a medium with refractive index μ, the optical path length through the medium differs from that in a vacuum. This change in path length ΔL is given by

$$\Delta L = \int \Delta \mu \, dl \qquad (2.16)$$

where l is the length of the medium in the direction of the wave propagation and $\Delta \mu$ is the change in refractive index. As discussed in section 2.1, the refractive index of a medium can be expressed as the sum of effects due to each species. If the jth component has relative refractive index of K_j, and density n_j, then to a first approximation

$$\mu - 1 = \sum K_j n_j. \qquad (2.17)$$

For the case when the medium is composed of neutral particles, positive ions and electrons (indicated by the subscripts n, i, and e respectively), equation (2.17) becomes

$$\mu - 1 = K_n n_n + K_i n_i + K_e n_e. \qquad (2.18)$$

For the case with no magnetic field, or for the case when the electric field of the electromagnetic wave is parallel to the magnetic field (called the O-mode), K_e can be found from the dispersion relation, equation (2.11). Consequently, for waves with angular frequency ω_0, the refractive index due to the electrons is given by $\mu_e = kc/\omega_0$, and so

$$\mu_e = [1 - (\omega_{pe}^2 / \omega_0^2)]^{1/2}. \qquad (2.19)$$

When $\omega_0 \gg \omega_{pe}$, the right-hand side of the equation can be approximated by $1 - 1/2(\omega_p^2/\omega_0^2)$ to give

$$K_e = - \frac{e^2}{8\pi^2 c^2 m_e \varepsilon_0} \lambda_0^2. \qquad (2.20)$$

Figure 2.12 shows the relationship between K_n, K_i and K_e and the wavelength of the electromagnetic wave. It can be seen that K_i and K_n have virtually no dependence on wavelength, whereas K_e depends linearly on λ_0^2.

The size of the fringe shift is given by equation (2.16) together with equation (2.17) or equation (2.18). Interferometers are widely used to measure

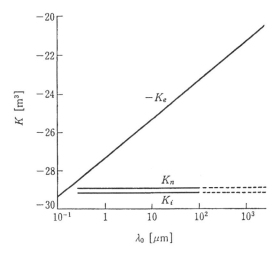

Figure 2.12. The dependencies of K_n, K_i and K_e [(from (2.12)] on the wavelength of the electromagnetic wave.

electron density in plasmas, and table 2.3 shows the value of the product of electron density n_e and measurement path l, which corresponds to one fringe shift (i.e. phase shift of $\Delta\phi = 2\pi$), for typical interferometer wavelengths.

Table 2.3. Values of $n_e l$ that produce one fringe shift for different lasers.

	Wavelength	$n_e l$ (m^{-2})
Ruby laser	694.3 nm	3.2×10^{21}
YAG laser	1.06 μm	2.1×10^{21}
CO$_2$ laser	10.6 μm	2.1×10^{20}
HCN laser	337 μm	6.6×10^{18}

Measurement of $\Delta\theta$. The polarization of linearly polarized electromagnetic waves passing through a medium changes if the medium possesses optical anisotropy. Gases, except for those composed of polar polyatomic molecules, usually do not have optical anisotropy, but magnetic fields sometimes can produce a strong anisotropy in a plasma. The direction of polarization of a wave with wavelength λ passing through a plasma with length l will change by an angle $\Delta\xi$ that is given by [5]

$$\Delta\xi = 2.6 \times 10^{-15} \int n_e B \, dl. \qquad (2.21)$$

In this expression, l has units of m, n_e has units of m^{-3} and B has units of tesla.

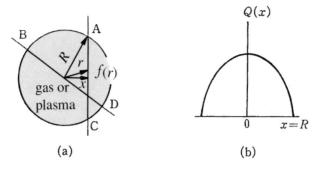

Figure 2.13. Principle of Abel inversion.

If the density can be determined by some other measurement technique, such as interferometry, then measurement of $\Delta\xi$ enables B to be determined. The usefulness of this method is found mainly in high-temperature plasma research, in devices with large magnetic fields such as tokomaks. In a tokamak discharge, it is necessary to measure the current density j, and this can be achieved by measuring B and then calculating j using $j = \nabla \times \boldsymbol{B}$. This technique has been developed into a sophisticated diagnostic method, called the *Faraday rotation method*, or *polarimetry*.

Abel inversion. In the transmission-based diagnostic methods described above, the value that is actually measured corresponds to the quantity integrated along the path of the laser beam inside the plasma. This can be a problem in the types of plasma that are studied in laboratories because these plasmas typically have large spatial variations in the plasma properties. The local value of each property cannot be obtained directly by transmission methods. In some circumstances, such as when the plasma cross-section is cylindrically symmetric, however, it is possible to indirectly derive the local value using a method known as Abel inversion [6]. In this section, the main features of Abel inversion will be outlined.

The properties of a plasma with a cylindrical cross-section can be represented in the manner shown in figure 2.13(a). For a quantity which reflects a property of a medium such as temperature or density, represented by $f(r)$, the line integral of $f(r)$ along the chord AC, represented by $Q(x)$, is given by

$$Q(x) = 2\int_x^R \frac{f(r)r\, dr}{(r^2 - x^2)^{1/2}}. \tag{2.22}$$

In this discussion, it is assumed that $f(r) = 0$ for $r > R$. If the position x of the chord AC is changed between 0 to R, the spatial distribution of Q can be measured. Such a profile is shown in figure 2.13(b). The local value of $f(r)$ then can be obtained from the measured values $Q(x)$ by inverting equation (2.22),

$$f(r) = -\frac{1}{\pi} \int_x^R \frac{Q'(x)dx}{(x^2 - r^2)^{1/2}}. \tag{2.23}$$

In this expression, $Q'(x)$ is the first derivative of Q with respect to x. This procedure, the derivation of $f(r)$ from $Q(x)$, is called Abel inversion.

Because $Q(x)$ usually is obtained as a discrete set of data, in most cases it is necessary to numerical integrate equation (2.23) to derive $f(r)$ from the measured values of $Q(x)$. However, Abel inversion is widely used for many measurement purposes in addition to the plasma and gas applications discussed here, and subroutines for numerical integration of equation (2.23) have existed for many years.

Although the plasma properties are not cylindrically symmetric, for the case when an $f(x,y)$ distribution can be assumed, a similar numerical integration can be performed. In this case, a measurement along only one direction, such as the x-axis measurement discussed above, is insufficient, and a further measurement is necessary, along another chord at some angle to AC, such as the chord BD shown in figure 2.13(a). This method is essentially the same as that used in many imaging diagnostics used in medical research, such as x-ray imaging, that are used to examine the human body. These techniques are collectively referred to as *tomography* or *computer tomography* (CT).

2.2.3 Refraction

In the same way that light passing through a lens or prism is refracted due to the difference of refractive index between glass and air, electromagnetic waves in a plasma or gas can be refracted due to a gradient in the refractive index of the medium. When the medium is a plasma consisting of neutral particles, ions and electrons, the gradient in refractive index $\nabla \mu$ is given, from equation (2.18), by

$$\nabla \mu = K_n \nabla n_n + K_i \nabla n_i = K_e \nabla n_e. \tag{2.24}$$

As discussed previously, and as shown in figure 2.12, K_n and K_i do not vary greatly with wavelength, while the wavelength dependence of K_e can be determined using equation (2.20).

There are two cases in which the refractive effect of a plasma on a gas becomes important for measurement purposes. The first of these is when the density gradients ∇n_j ($j = n,i,e$) change very sharply in a thin layer, such as occurs in a shock wave. Refraction of light at the density gradients can be used to determine the position of the shock wave. Examples of techniques that use this effect are the schlieren method and shadowgraphy. In addition, the refractive index of a plasma has to be assessed so that the amount of refraction of a laser beam passing through the plasma can be determined. This has to be done in order to ensure that the refractive effect is small enough for transmission measurements to be feasible.

Schlieren method. Figure 2.14 shows an example of an apparatus for

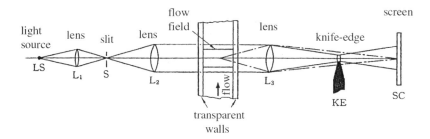

Figure 2.14. Example of an apparatus for measurement using the schlieren method.

observing a shock wave propagating through air in a tube [7]. Light from the source is focused onto a slit using a cylindrical lens L_1 to produce a line-shaped beam. The beam then is collimated using another lens L_2, directed across the flow, and focused to the position of the knife-edge using lens L_3. The light then is projected onto a screen that is placed at the focal plane of the flow by the lens L_3. The knife-edge KE is a sharp-edged plate that is positioned just below the focused point of the beam.

In this apparatus, the light from the source will uniformly illuminate the screen if there are no density gradients in the gas inside the flow tube. However, when there are density gradients in the direction of the flow, there will be gradients in the refractive index of the gas inside the flow tube, and so light in the parallel beam will be refracted (i.e. bent) as it passes through the flow. The refracted part of the light is obstructed by the knife-edge. Therefore, the screen will be illuminated uniformly except for regions that depend on light passing through the part of the flow that has a density gradient. In these regions of the screen, the light will become brighter or darker, depending on the phase of the light reaching the screen at that point. In this arrangement, only density gradients in the direction of the slit and the knife-edge can be observed. By changing the orientation of these components, refractive index changes in other directions also can be observed.

The angle through which the light is refracted, ε, depends linearly on the gradient in the refractive index. Hence,

$$\varepsilon \propto \nabla \mu. \qquad (2.25)$$

For the case of a shock wave travelling through the atmosphere, n_e and n_i in equation (2.18) are zero, and so equation (2.25) can be written as

$$\varepsilon \propto K_n \nabla n_n. \qquad (2.26)$$

Hence, ε depends linearly on the density gradient of neutral particles in the gas. It is possible to view this effect simply by looking at the screen by eye, but usually some form of two-dimensional imaging device is used to detect the image either photographically or electronically. For this type of measurement, a

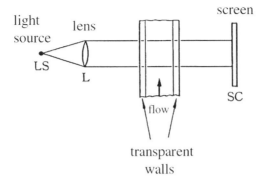

Figure 2.15. Example of an apparatus for measurement using shadowgraphy.

pulsed light source such as a Q-switched ruby laser is used, and triggered in phase with the shock wave.

Shadowgraphy. Figure 2.15 shows an example of an apparatus used to measure the propagation of a shock wave using the technique of shadowgraphy [7]. Light from a point source is passed through a lens and is recorded directly on the screen. It is not necessary for the beam to be perfectly collimated. When there is no flow in the tube, the screen SC is uniformly illuminated but when there is a flow, and hence a density gradient due to this flow, the brightness on the screen B will be changed by an amount ΔB given by

$$\Delta B \propto \left(\frac{\partial^2}{\partial y^2} + \frac{\partial^2}{\partial z^2} \right) \mu. \tag{2.27}$$

In this equation, y and z are the position coordinates on the screen surface. Because the second derivative of μ is recorded, this provides a way of observing sharp density changes. In shadowgraphy, the detector is the same kind of photographic or electrical recording device used for the schlieren method.

2.2.4 Scattering

When a medium such as a plasma or gas contains only a single type of particle, it is relatively easy to interpret the light scattered from the medium. However, scattering from a medium that contains several different species is more complicated. The angular and spectral distribution of the scattered light, and in some cases the response of the scattering to changes in the wavelength of the laser radiation, are used to investigate the contribution of the different scattering phenomena.

The scattered light from a gas or plasma arises because of the phenomena discussed in section 2.1: Rayleigh scattering, Raman scattering, Mie scattering and Thomson scattering. The amount of scattering that occurs depends both on the properties of the medium, such as the type of species present and the degree of ionization, and the properties of the laser, such as its wavelength, spectral width and output power. These types of scattering types are discussed below.

Thomson scattering. As discussed in section 2.1.1, Thomson scattering from plasmas can be thought of as only being due to electrons, because electrons are accelerated by the laser electric field by a much larger amount than the ions, simply due to their much smaller mass. The Thomson scattering cross-section for scattering from a single electron was given earlier, by equations (2.3)–(2.6). The scattering due to a group of electrons can be derived using a combination of Maxwell's electromagnetic theory and the Vlasov equation from plasma theory [1]. The results of this derivation are shown below.

For the case when the electron density has both spatial and time dependence, the electron density $n_e(r,t)$ can be considered to consist of a time-averaged component $\bar{n}_e(r)$ and a fluctuating component $\tilde{n}_e(r,t)$, and so can be written as

$$n_e(r,t) = \bar{n}_e(r) + \tilde{n}_e(r,t) \qquad (2.28)$$

The differential scattering cross-section $\sigma(\lambda,\theta)$ depends only on the fluctuating component $\tilde{n}_e(r,t)$. In the lower limit when $\tilde{n}_e(r,t)$ is due only to thermal fluctuations, $\sigma(\lambda,\theta)$ can be expressed as a component due to the thermal motion of the electrons, $\sigma_e(\lambda,\theta)$ and a component due to the effect on the electrons of fluctuations in the ion thermal motion, $\sigma_i(\lambda,\theta)$. Thus, $\sigma(\lambda,\theta)$ can be written as

$$\sigma(\lambda,\theta) = \sigma_e(\lambda,\theta) + \sigma_i(\lambda,\theta) \qquad (2.29)$$

If the scattered wavenumber and frequency are written as follows (cf. figure 2.3),

$$\begin{aligned} k &= k_i - k_s \\ \Delta\omega &= \omega_i - \omega_s \end{aligned} \qquad (2.30)$$

then for the case when $k\lambda_D \gg 1$, only the thermal electron part of the cross-section, $\sigma_e(\lambda,\theta)$, is important. Here, $|k| = k$ is the magnitude of the differential wavevector and λ_D is the Debye length given by equation (1.2). This criterion implies that the scattering wavelength is much smaller than the Debye length. This situation, in which the scattering is only due to the thermal motion of the electrons, is called *incoherent scattering*.

If the electron velocity distribution function is a Maxwellian distribution, then σ becomes

$$\sigma = \sigma_e = \frac{Kn_e c}{2\omega_i \sin(\theta/2)\sqrt{\pi}v_{th}} \exp\left\{-\frac{1}{v_{th}^2}\left[\frac{\Delta\omega_D c}{2\omega_i \sin(\theta/2)}\right]\right\}. \qquad (2.31)$$

In this expression, $\Delta\omega_D = k \cdot v$ and K is a constant. This equation shows that the scattered spectrum is broadened from that of the incident light by an amount related to the thermal velocity distribution of the electrons, and so the electron temperature can be determined by measurement of the scattered light spectrum.

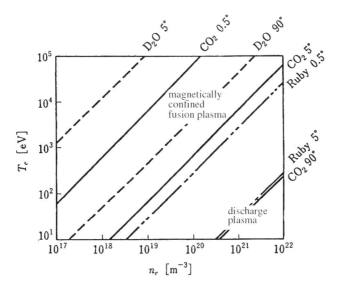

Figure 2.16. Range of plasma conditions plotted on n_e–T_e coordinates with the $k\lambda_D = 1$ line shown for typical lasers and scattering systems. Collective scattering occurs for the region below and to the right of the lines. The wavelengths are 694 nm for ruby laser, 10.6 μm for CO_2 laser, and 385 μm for D_2O laser.

In addition, as can be seen by considering equation (2.2) and equation (2.31), the scattered light intensity is proportional to the electron density, and so electron density can be determined experimentally by measuring the Thomson scattered light intensity.

If $k\lambda_D < 1$, which means that the scattered wavelength is longer than the Debye length, the scattering that occurs is due to the collective motion of the electrons that surround and follow the ion motion. This regime is called *collective scattering*. In this case, σ_e and σ_i have the same order of magnitude, and the intensity is related to the effective charge $\overline{Z}$ of the particles in the discharge, and the electron and ion temperatures T_e and T_i.

The above discussion is about the scattering cross-section in the limit of thermal fluctuations. Scattering also can be observed due to other, i.e. non-thermal, fluctuations. When waves are induced in a plasma, and when those waves have an amplitude that is sufficient to cause nonlinear wave effects, scattering provides a means of detecting these wave effects. However, even in this case, when the wavelength of the induced wave is smaller than the Debye length, the perturbations due to the wave are very quickly damped by a mechanism called *Laudau damping* [3], and for $k\lambda_D \ll 1$, only thermal fluctuations exist.

As should be clear from this discussion, $k\lambda_D = 1$ is the boundary between the two regions of Thomson scattering. Figure 2.16 shows plasma regimes plotted on a graph with n_e and T_e as coordinates, with the $k\lambda_D = 1$ boundary being shown for typical laser sources. For the high-temperature plasmas that are used in fusion research, ruby laser scattering detected at 90° is incoherent scattering, and this can be used as a diagnostic method for n_e and T_e. Also for these plasmas, scattering from microwaves is collective scattering, and this can be used to determine ion temperature or plasma waves/turbulence. In addition to these fusion plasma examples, incoherent Thomson scattering has been applied recently in glow discharges as a method for measuring electron properties.

Mie and Rayleigh scattering. The optimum laser for most Mie scattering experiments is a high-power, short-wavelength cw laser, such as an Ar-ion laser. Measurements of instantaneous phenomena are possible by using pulsed lasers. When the particle density is small, scattered light from each particle can be observed discretely, but in most cases the observed light intensity corresponds to a value averaged for many particles.

Mie scattering is used for a variety of purposes. One example of an application is the measurement of gas flow velocity, described in chapter 6, in which scattered temporal light trains from small particles in different locations are observed. Another example is the detection of small particles generated during laser processing applications. In this case, information about both particle density and particle size is desired. Because the Mie scattering intensity depends on both these quantities, it is not possible to determine both quantities from a single set of scattering measurements, such as those shown in figure 2.4. However, by performing two different scattering experiments, such as using two different wavelengths, or beams with different polarizations, these two quantities can be determined. If there is a distribution of particle sizes, then some assumption about that distribution has to be made in order to interpret the measured results.

It can be seen from figure 2.5 that the detection limit decreases rapidly as the particle size decreases. Consequently, it is difficult to measure nanometre-size clusters of particles using Mie scattering. In this case, TOF laser mass spectroscopy can be used. Larger particles, of micrometre-order size, can be detected using in-line holography, which uses diffraction and interference between the particles themselves to generate the signal [8].

Except for the fact that the scattered wavelength is the same as the laser wavelength, the optical arrangements for Mie and Rayleigh scattering experiments have few common features. Because the Rayleigh scattering intensity is usually very weak, pulsed high-power lasers are used. In addition, spectrometers are used to filter out signals at different wavelengths. In this sense, the optical arrangement for Rayleigh scattering is similar to those used for Thomson scattering, Raman scattering and fluorescence scattering.

Applications of Rayleigh scattering include plasma diagnostics, combustion gas measurements, and environmental monitoring. These are described in chapters 4, 5 and 9, respectively.

Raman scattering. Raman spectroscopy is a technique that has progressed greatly since the development of lasers, because measurements with conventional light sources were virtually impossible due to the small Raman

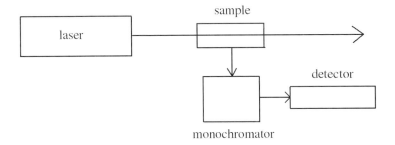

Figure 2.17. Schematic diagram of an apparatus for laser Raman spectroscopy.

scattering cross-section. Figure 2.17 shows a typical experimental arrangement for laser Raman spectroscopy. Photon-counting devices usually are used for the detection system, and reduction of the stray light is very important to improve the detection limit. The most useful excitation source is the cw argon-ion laser. The main advantage of laser Raman spectroscopy is that many kinds of molecules can be identified simultaneously using a single fixed-frequency laser.

When the excitation laser light is very intense, stimulated Raman scattering at a Raman-shifted frequency can be observed. This is a kind of nonlinear parametric process, and is used as the basis of a frequency converter called a Raman shifter that is used to change the frequency of light from coherent light sources. Raman shifters are very simple and efficient converters, especially when used with high-power pulsed lasers in the visible and UV regions. Details of these devices are given in section 3.2.4.

Various kinds of nonlinear Raman spectroscopy have been developed for

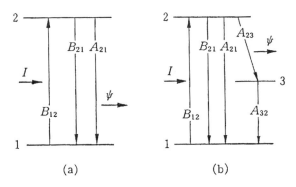

Figure 2.18. Energy level diagram showing resonant absorption and fluorescence. Excitation is by incident laser light that has intensity I and photon energy that matches the energy difference between levels 1 and 2. The fluorescence light ψ is observed. The transition probabilities A and B are discussed in the text. (*a*) shows a two-level system while (*b*) shows a three-level system.

the detection of particles. Most of these methods have better sensitivity than conventional Raman spectroscopy. The most useful techniques are CARS (coherent anti-Stokes Raman spectroscopy) and DFWM (degenerative four-wave mixing). Details of these methods are given in section 5.1.2.

Fluorescence (resonant scattering). Fluorescence is observed when atoms and molecules are excited to higher energy levels and then radiatively relax. The method based on this phenomenon, laser-induced fluorescence (LIF), is the most widely used laser diagnostic technique in almost all of the research fields discussed in this book. Extremely high sensitivity can be achieved when the LIF scheme uses electronic transitions with visible and ultraviolet wavelengths.

LIF schemes can be divided into the two basic types of spectroscopic schemes shown in figure 2.18. In the first scheme, only two levels are used, connected by a single transition, and the excitation and fluorescence wavelengths are the same. This type of fluorescence is sometimes called resonant scattering. In the second type of scheme, more than two levels are involved, and the excitation and fluorescence wavelengths are different.

A typical arrangement for LIF measurements is shown in figure 2.19(a). The fluorescence light is collected by a lens and detected by a photomultiplier. For the two-level case of figure 2.18(a), the LIF signal intensity is given by [9]

$$I_F = \eta \frac{\Omega}{4\pi} \frac{g_2}{g_1 + g_2} \frac{S}{1+S} V \frac{N}{\tau}. \qquad (2.32)$$

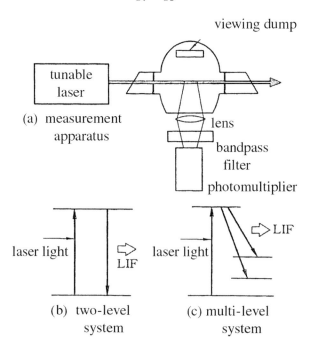

Figure 2.19. Arrangement for LIF spectroscopy.

In this expression, η is the detection efficiency of the detection system, Ω is the solid angle through which the signal is detected, g_1 and g_2 are the degeneracies of the two levels, V is the measurement volume, τ is the fluorescence lifetime and N is the density of the absorbing species. S is a parameter called the saturation parameter, given by

$$S = I(\nu_0) \frac{g_1 + g_2}{g_1} \frac{c^2}{8\pi h \nu_0^3} g(\nu_0) \tag{2.33}$$

where $I(\nu_0)$ is the incident laser power and $g(\nu_0)$ is a function indicating the spectral broadening of the line.

It can be seen from equation (2.32) that I_F is directly proportional to the particle density N. If the measurement system is calibrated by some method, the absolute value of N can be determined from the measured signal. In addition, when the excitation is weak, and hence S is small, I_F is proportional to the laser power, while when the laser power I is sufficiently large, S becomes large and I_F saturates. Once the signal has saturated, the signal does not increase further, no matter how much higher the laser power is increased. The most sensitive method to obtain LIF signals is to use a cw tunable laser with output power close to the saturation power. In this case, excitation and relaxation occur repeatedly, with a period corresponding to the fluorescence lifetime, and a single particle can emit many photons per second. This process allows LIF to be very sensitive.

The detection limit for LIF measurements using a two-level LIF process, in practice, is determined mainly by stray light because the excitation and fluorescence transitions have the same wavelength. When the measurement is made in vacuum conditions, detection of 10^8–10^9 particles/m^3 is possible by suppressing the stray light. For measurements during laser processing applications, in the presence of background gas and background emission, detection sensitivity of the order of 10^{12}–10^{13} particles/m^3 is feasible if the background emission is relatively weak.

For LIF measurements using a multi-level LIF scheme (figure 2.18(b)), the detection and excitation wavelengths are different, and stray light can be eliminated by using an optical filter in front of the detector. However, compared with a two-level LIF scheme, the signal is reduced because the fluorescence is shared among several transitions. Furthermore, if there are levels in the system with long lifetimes, particles can become 'trapped' on these levels and so cannot be re-excited. Hence, although stray light is much less of a problem, it is not necessarily true that the detection sensitivity for multi-level LIF schemes is higher than those of simple two-level schemes.

When a pulsed laser is used for excitation instead of a cw laser, the number of LIF photons that can be produced is limited due to the short pulse length, and averaging of the signal over many laser pulses becomes important. Spatial distributions often can be measured instantaneously using the LIF method combined with one-dimensional and two-dimensional detectors. Although LIF techniques have many advantages, it should be noted that they are difficult to apply in high-pressure environments when quenching due to particle collisions is severe.

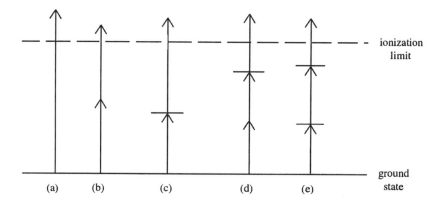

Figure 2.20. Various schemes for photo-ionization: (*a*) single-photon ionization, (*b*) two-photon ionization, (*c*) two-step ionization, (*d*) two-photon excited ionization, and (*e*) three-step ionization.

2.2.5 Photo-Ionization

Photo-ionization, described in section 2.1.5, can be used to measure a variety of species present in gases, and highly selective ionization is possible by using a tunable laser. *Resonance ionization spectroscopy* (RIS), like LIF, is an extremely sensitive particle detection method. By combining a high-intensity laser that can produce 100% ionization with a high-efficiency ion detector, detection at the level of single particles is feasible.

The photo-ionization spectrum is usually continuous, and so it is difficult to identify different species from the ionization spectrum. Hence, direct ionization schemes such as those shown in figures 2.20(*a*) and (*b*) cannot be used to differentiate between species. Figures 2.20(*c*)–(*e*) show various schemes that use multi-step ionization, so that different species can be detected. In these schemes, the ionization probability depends strongly on the excitation wavelength. Figures 2.20(*b*) and (*d*) show examples of multi-photon schemes. In general, resonant ionization spectroscopy has very high sensitivity, because ion counting with a high signal-to-noise ratio can be achieved.

In order to produce a high degree of ionization, laser power of kW–MW order is necessary, and so pulsed lasers usually are used as the laser source in photo-ionization measurements. Because the laser power is very high, ionization due to multiple photon absorption from the ground state can occur, as can simultaneous ionization of impurity species. These effects reduce the signal-to-noise level in a real experiment. It is technically possible to remove the signals due to ionization of other species by using a mass filter, but this requires a separate high vacuum system and, also, the *in situ* aspect of the measurement is lost. In addition, it should be noted that RIS methods are difficult to apply in environments that produce ions efficiently by themselves, such as plasmas.

The main merit of RIS methods, compared with LIF methods, is that

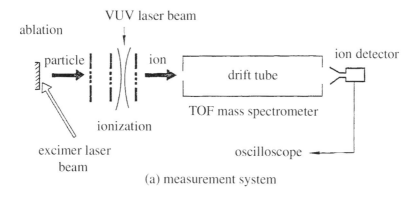

(a) measurement system

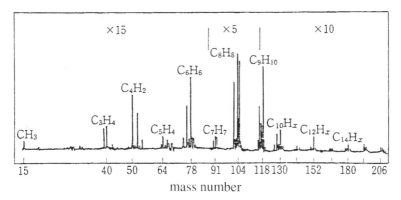

(b) mass spectrum of particles generated by ablation of a polymer

Figure 2.21. Apparatus and example spectrum for laser mass spectroscopy [8].

species that do not fluoresce can be detected. Another advantage is that, because the signal is obtained simply by placing a collection electrode in the measurement region, signals can be obtained from media that have high background light emission levels.

In *laser mass spectroscopy* (LMS), a laser is used to non-selectively ionize particles, and a mass spectrometer then is used to detect the ionized species. If a pulsed laser is used as an ionization laser, a time-of-flight (TOF) mass spectrometer using a simple drift tube can be used to differentiate between species. An example of a laser mass spectrometer is shown in figure 2.21(a), together with a typical result in figure 2.21(b). The main advantage of laser mass spectrometry is that a large number of different species can be detected simultaneously. In addition to atomic and molecular species, clusters of small particles also can be measured.

54   *Basic Principles*

Optogalvanic spectroscopy (OGS) is a type of selective ionization method, and is described in sections 4.3.1 and 8.2.4.

2.3 Spectral Profile Measurements [10]

There are many different factors that affect the spectral line profiles of atoms, molecules and ions. These factors include the processes called natural broadening, pressure broadening, Stark broadening, Zeeman broadening and Doppler broadening. Each of these broadening mechanisms reflects some property of the emitting species, such as its transition probability, collision frequency, thermal velocity, and the local electric and magnetic fields. If it is possible to identify and measure the spectral line profiles of atoms and molecules, these properties can be determined.

Spectral lines usually have widths of the order of picometres, and instruments such as Fabry–Perot interferometers can be used to measure and analyse emission spectra. Although this form of passive spectroscopy sometimes is feasible, spectral profile measurements have become much more useful since the development of lasers, because excitation of specific transitions and observation of spectral lines with good spectral and spatial resolution is possible. In addition, many new spectroscopic techniques have been developed to take advantage of the narrow spectral width and tunability of lasers such as dye lasers.

This section contains a summary of the main features of spectral profile measurements. The factors that influence spectral profiles are described in section 2.3.1, some examples of profile widths are presented in section 2.3.2, and a discussion of the performance required from lasers and detection systems for different measurement techniques is contained in section 2.3.3.

2.3.1 Summary of Line Broadening Mechanisms

There are many different factors that affect the widths of spectral lines of atoms and molecules in gases and plasmas. The most important factors are explained briefly below.

Natural broadening due to relaxation processes. In general, transitions from energy levels that possess a finite lifetime τ will have a Lorenztian lineshape that is characterized by a width $\Delta\nu = (1/2\pi\tau)$. Many different factors influence the lifetime τ, but when spontaneous emission completely determines the lifetime, the width of the spectral line is called the *natural width*. For cases when the pressure is high, collisions play a major role in depopulation of the energy level, and the lifetime τ becomes equal to the inverse of the collision frequency. This causes the spectral profile to be broader than the natural width, and the width in this case is called the *pressure-broadened width*.

Stark broadening. Energy levels of bound electrons in atoms and molecules are affected by the presence of an electric field. This is called the *Stark effect*, and can lead to broadening of the spectral profile that is linear with the electric field (first-order Stark effect), or a shift in the transition wavelength (second-order Stark effect). For particles in the sheath region of a plasma, where there is a relatively strong electric field ($E > 10$ V mm^{-1}), both first-order and second-order effects can be observed. For particles in the bulk plasma, where the electric field is very small, the microfields caused by fluctuations in the plasma density produce first-order Stark broadening. This latter effect can be used as a method to estimate the electron density. Also, there can be a mixing of energy states due to Stark effects, and this can have a complicated effect on the spectral profile. It is sometimes possible to measure the electric field by observing these effects.

Measurement methods based on these effects are described in section 4.3.1.

Zeeman broadening. The presence of a magnetic field also affects the energy of bound electrons in atoms and molecules. This is called the *Zeeman effect*. This kind of effect can occur due to the effect of not only an applied magnetic field, but also by the magnetic field produced by currents flowing in a plasma. In a fusion plasma, this latter effect is used to determine the plasma current, by measuring the splitting of spectral lines and then calculating the magnetic field. This technique is explained in section 4.2.3.

In the kinds of glow discharge plasmas used for industrial processing purposes, spectral broadening due to Zeeman effects is rarely important because, with the exception of the ECR discharge, most processing plasmas do not have an applied magnetic field that is strong enough to affect spectral lineshapes.

Doppler broadening. A particle that has a resonant frequency v_0 will have that resonant frequency shifted if it is moving due to the *Doppler effect*. When the particle is moving at a velocity v, the frequency is shifted by $(v/c)v_0$. The velocity v that determined the shifted frequency is the velocity in the direction in which the particle is observed. Species that have a Maxwellian velocity distribution due to thermal motion will have a spectrum with a Gaussian shape. The full width at half-maximum Δv_D of the spectrum is given by

$$\Delta v_D = \frac{2v_0}{c} \sqrt{\frac{2kT \ln 2}{m}} \qquad (2.34)$$

where m is the particle mass, and T is the temperature. This width is called the *Doppler width*.

Doppler broadening of spectral lines is due to the Doppler motion of individual particles, and is a form of *inhomogeneous broadening*. The other effects discussed above, such as natural broadening, pressure broadening and Stark broadening, are examples of *homogeneous broadening*.

Fine structure broadening. Splitting of spectral lines due to fine structure sometimes can be an important factor in determining the overall spectral width. For atoms, there can be fine shifts due to the presence of different isotopes of

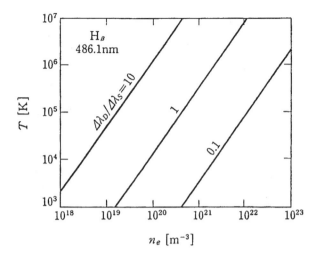

Figure 2.22. Ratio of Doppler width $\Delta\lambda_D$ and Stark width $\Delta\lambda_s$ for the hydrogen H_β transition for a wide range of plasma conditions.

the same species, as well as fine structure due to differences in orbital and spin angular momentum. For molecules, in addition to these effects, the presence of vibrational and rotational energy levels also has to be considered. Transitions from excited rotational levels that exist in the ground state are called *hot bands*, and the spectral profiles of molecules often depend on temperature in a complicated way.

2.3.2 Examples of Spectral Widths

In this section, examples of typical widths due to the broadening mechanisms discussed above will be presented. The Doppler width of the 302 nm resonance line of iron atoms at 1000 K is about 0.3 GHz (0.9 pm) and the Doppler width of the sodium D_2 lines at 500 K is about 1.7 GHz (2 pm). Compared with these widths, the natural widths of transitions from energy levels in atoms and molecules are very small, about 10 MHz, because these levels usually have relatively short lifetimes of the order of 10 ns. For the sodium D_2 lines, the width due to pressure broadening in argon gas is about 30 MHz/Torr, and about 150 MHz/Torr for collisions between the sodium atoms themselves. Consequently, for pressures below a few Torr, Doppler broadening is much more significant than natural or pressure broadening.

The Stark width of the hydrogen H_β line in a plasma with $n_e \sim 10^{20}\,\text{m}^{-2}$ is 42 pm, and so must be considered for high-density plasmas. Glow discharge plasmas, however, have much lower electron density, and for these discharges, the Stark broadening is small. In these plasmas, it is mainly Doppler broadening

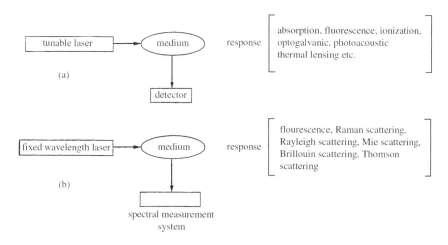

Figure 2.23. Schematic diagram showing arrangements for spectral measurements using (*a*) excitation methods and (*b*) scattering methods.

that determines the spectral line profile. For example, hydrogen atoms in a glow discharge with gas temperature of 1000 K have a Doppler width of about 12 pm. The relative sizes of Stark and Doppler widths are shown in figure 2.22. For the case of an ECR plasma, which uses absorption of 2.45 GHz microwaves at the resonant magnetic field strength of 87.5 mT to generate the plasma, Zeeman splitting has to be considered. The Zeeman splitting of iron atoms in such a discharge is about 0.6 pm, while the Doppler width at $T \sim 1000$ K is about 0.9 pm. Hence, the total spectral width is a combination of both these effects.

Because effects such as fine structure, isotope shifts and molecular rotational levels lead to broadening of the order of 1 cm^{-1} (30 GHz), these effects can have a direct and large influence on the spectral profile.

The temperature of atoms and molecules can be measured either from the Doppler width or from the distribution of population among excited levels. The latter method requires the assumption of a Boltzmann distribution among the sub-levels, and it also is necessary to know the transition probability for each transition from each individual level. These methods will be discussed in section 4.1.

2.3.3 Spectral Profile Measurement Techniques

Methods for measuring spectral profiles can be divided into the two categories shown in figure 2.23: excitation methods and scattering methods. In the first category, the wavelength of a tunable laser is scanned, and the signal intensity is measured as a function of wavelength. A spectrum obtained in this way is called an *excitation spectrum*. In all excitation techniques, the amount of light absorption that occurs in the medium is measured, but not all methods measure the absorption directly. Most methods measure some other effect (emitted light, ionization, sound emission, change in refractive index etc.) and then indirectly determine the amount of absorption. The reason for this is that when the absorption is directly measured, the change in laser intensity due to the absorption has to be larger than fluctuations in the laser intensity, which even when small are of the order of 1%. When indirect methods are used, however, absorption levels much lower than this can be detected.

In the second category of measurement methods, which is comprised of scattering techniques such as Raman and Thomson scattering, a fixed frequency laser usually is used. In these measurement methods, the spectral resolution is determined by the spectrometer that is used to make the measurement rather than the laser spectral width, as is the case for excitation methods. In addition, in scattering measurements, the signal is detected through a narrow spectrometer slit and the measured light is spectrally dispersed before detection. This necessarily results in a smaller signal and hence a decrease in sensitivity.

Fluorescence spectroscopy, which is widely used in measurements of processing plasmas, can be performed using methods of either category. For measurements of fine structure with widths of less than 1 cm^{-1}, excitation methods must be used. Molecular spectra, however, have broader profiles and can be measured using both scattering and excitation methods.

When excitation methods are used, the spectral width of the laser often is comparable to the width of the profile to be measured, and so this affects the measurement. However, the spectral shape of the laser light can be measured separately, and so in many cases it is possible to determine the true spectral line shape by de-convolution. If the laser is operated in a single mode, but has a range of modes spread over a 100–300 MHz frequency range, it often is possible to ignore this structure because this is still much narrower than most profiles that are measured. However, if saturation effects are significant, this fine structure may lead to extra broadening. For the cases when the spectral profile is distorted due to saturation, analysis based on a rate equation model of the absorption processes must be used.

Good signal-to-noise ratios are obtainable for spectral profile measurements as long as a sufficiently long time can be used for the measurement. However, because the laser wavelength is scanned over time in a spectral profile measurement, spectral measurements take a much longer time than density measurements, and there is often a limit on the amount of time that can be used for the measurement. The laser scanning rate and the detector averaging time determine the total time needed for a particular measurement, and if this becomes too long, the spectral resolution might be degraded, or uncertainty might be introduced into the measurement. One way to improve the signal-to-noise ratio is to increase the laser intensity in order to increase the signal, but if the laser intensity is high enough to cause saturation, then the spectral profile will become distorted.

REFERENCES

[2.1] Evans D E and Katzenstein J 1969 *Rep. Prog. Phys.* **32** 207
[2.2] Hulst H C 1957 *Light Scattering by Small Particles* (Wiley)
[2.3] Stix T 1962 *Theory of Plasma Waves* (McGraw-Hill)
[2.4] Sadeghi N 1999 *Proc. 9th Int. Symp. Laser-Aided Plasma Diagnostics (Lake Tahoe, USA, 26 September – 1 October)* p 383
[2.5] van Milligan Ph, Soltwisch H and Lopes Cardozo N J 1991 *Nucl. Fusion*, **31** 309
[2.6] Nestor H and Olsen H N 1960 *SIAM Rev.* **2** 200
[2.7] Liepmann H W and Roshko A 1960 *Elements of Gas Dynamics* (Wiley)
[2.8] Thomson B J 1965 *Japan J. Appl. Phys.* **4** Suppl. 4-1, 302
[2.9] Feldmann D *et al* 1987 *Appl. Phys.* B **44** 8
[2.10] Muraoka K, Honda C, Kim J J, Yamagata Y and Akazaki M 1989 *Rev. Laser Eng.* **17** 546 (in Japanese)

Chapter 3

Hardware for Laser Measurements

3.1 Lasers

Since laser oscillation was first obtained using a ruby crystal, and the ruby laser was developed in 1960, many different lasers have been developed, with wavelengths covering a very wide range, based on a variety of different materials. This section contains a brief summary of important background information about lasers, and also an overview of laser systems that are commercially produced. Because many types of lasers now are commercially available, it is rarely necessary for a user to build a laser system. However, in order to understand and use laser sources, a certain basic amount of knowledge is useful [1–4].

3.1.1 Overview of Laser Systems

All laser systems are based on transitions between energy levels in atoms or molecules. Figure 3.1 shows atomic and molecular transitions that involve two levels only. When a transition between the levels occurs, energy is exchanged between the particle and its surroundings. This exchange of energy can be divided into two distinct cases. In one case, called a radiative transition, the change in energy is in the form of an electromagnetic wave (i.e. a photon). In the other case, called a non-radiative transition, the change in energy is thermal or vibrational (i.e. a phonon).

There are three types of radiative transitions: spontaneous emission, absorption, and induced emission. The transition probabilities per unit time, shown in figure 3.1, are given by A_{21}, $\rho(\nu)B_{12}$, and $\rho(\nu)B_{21}$, respectively. A and B are called Einstein coefficients, $\rho(\nu)$ is the incident light energy density, and the relationship between the frequency ν of the absorbed or emitted light, and the energy difference between the levels ΔE is given by $\Delta E = h\nu$, where h is Planck's constant.

The interaction of light with matter can be considered using quantum mechanical theory, based on a perturbation approach. In this theory, the probability of emission and absorption depend linearly on $\rho(\nu)$, from the first-

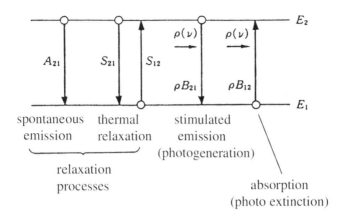

Figure 3.1. Energy exchange mechanisms in a two-level system.

order term of the perturbation, and nonlinear effects come from higher-order terms. For the case when the energy levels are non-degenerate, the relationship between the Einstein coefficients is given by

$$A_{21} = \frac{8\pi h \nu^3}{c^3} B_{21}, \quad B_{21} = B_{12} \qquad (3.1)$$

where c is the speed of light.

Spontaneous emission is one of the relaxation processes by which a medium can return to thermal equilibrium. Spontaneous emission often is called fluorescence. The fluorescence lifetime τ, also called the relaxation time, is given by $\tau = 1/A_{21}$. Non-radiative relaxation processes also can occur, and these are shown in figure 3.1 with probabilities given by S_1 and S_2. The induced emission probability B_{21} is proportional to the spontaneous emission probability A_{21}, and so systems that are dominated by non-radiative transitions rarely are suitable for efficient generation of laser radiation.

Energy levels in atoms and molecules are always broadened to some extent, and the mian sources of broadening were described briefly in section 2.3. For example, the width of spectral lines of atoms or molecules in a gas are usually Doppler-broadened by the thermal motion of the particles, and the emission lines have Gaussian spectral profiles. When the upper level of the transition has a finite relaxation time τ, the spectrum is broadened by the uncertainty principle and the spectral profile is a Lorentzian function, with a half-width of $\Delta \nu = 1/2\pi\tau$. Hence, as τ becomes shorter, the width becomes broader. This form of broadening, due to the finite fluorescence lifetime, is called *natural broadening*, and the resulting width is called the *natural width* of the line. In most cases, the natural width is so small that it is buried within the Doppler width. For example, a lifetime of $\tau = 10$ ns corresponds to a natural

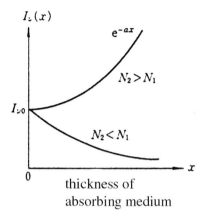

Figure 3.2. Spatial dependence of light intensity for light passing through a medium. The light is amplified or reduced, depending on the population distribution.

width of the order of 10 MHz, much smaller than typical Doppler widths. However, when the gas pressure becomes extremely high, so that the relaxation time due to collisions is very short, the natural width can exceed the Doppler width. This is called *pressure broadening*. In liquids and solids, the close proximity of particles perturbs the energy levels, and the spectral widths of transitions are much broader than those of gases.

The probability of absorption on a given transition is given by $B_{12}\rho_v g(\nu)$, where $g(\nu)$ is the normalized spectral distribution function, and ρ_v is the incident monochromatic light density. Absorption and stimulated emission are essentially the same phenomena, occurring in opposite directions, and as can be seen from equation (3.1), the transition probabilities for both phenomena are the same. For the case of a two-level system such as that shown in figure 3.1, absorption and stimulated emission occur simultaneously, and energy of $h\nu$ is exchanged in both directions. In this case, if other relaxation processes can be neglected,

$$\frac{d\rho_v}{dt} = h\nu(B_{21}N_2 - B_{12}N_1)\rho_v g(\nu) \tag{3.2}$$

This expression can be rewritten in terms of the flux of light $I_v = c\rho_v$, by writing $B = B_{21} = B_{12}$, from equation (3.1), and using $x = ct$. Equation (3.2) then becomes

$$\frac{dI_v}{dx} = \frac{h\nu}{c} B I_v g(\nu)(N_2 - N_1). \tag{3.3}$$

It also is possible to write this in terms of an absorption coefficient $\alpha(\nu)$. The change in intensity I_{v0} of a planar light beam passing through an absorbing medium is given by

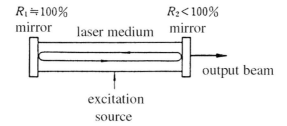

Figure 3.3. Schematic diagram of a Fabry–Perot laser resonator.

$$\frac{dI_v}{dx} = -\alpha I_v \quad \text{or} \quad I_v = I_{v0}\exp(-\alpha x). \tag{3.4}$$

Comparing equation (3.3) and equation (3.4), it can be seen that the absorption coefficient is given by

$$\alpha(v) = -hv\, B\, g(v)(N_2 - N_1). \tag{3.5}$$

Figure 3.2 shows $I_v(x)$ plotted as a function of x. In most cases, $N_2 < N_1$, α is positive, and the light intensity decreases as the beam passes through the medium. Conversely, as can be seen in the figure, for the case of $N_2 > N_1$, α is negative, and the light intensity increases. This situation, in which $N_2 > N_1$, is called *population inversion*. The range of wavelengths over which amplification occurs is determined from the spectral line profile $g(v)$, as indicated by equation (3.5).

To obtain laser oscillation, some kind of resonant cavity is required to provide feedback of the light to the laser medium. In most cases, the laser cavity is a *Fabry–Perot resonator*, which consists of two mirrors that are aligned to be parallel. Such a cavity is shown in figure 3.3. One mirror, called the output coupler, has reflectivity less than 100%, and light leaves the resonator through this coupler. This light forms the output laser beam.

Laser oscillators have a threshold for oscillation. This threshold condition is that the loop gain is equal to 1, where the loop gain is defined as the amplification of light produced by one round trip of the oscillator. This condition is the same as in electronic oscillators. When the two mirrors have reflectivity of R_1 and R_2, and the length of the resonant cavity is L, this oscillation condition can be written as

$$R_1 R_2 \exp(-2\alpha L) = 1. \tag{3.6}$$

Consequently, the amplification necessary for achieving oscillation is

$$\alpha = \frac{\ln(R_1 R_2)}{2L} = -\frac{1}{c\tau_c}. \tag{3.7}$$

In this expression, τ_c is the lifetime of the light inside the resonator. τ_c is related to the Q factor of the cavity, $Q = \omega_0 \tau_c$, where ω_0 is the resonant angular

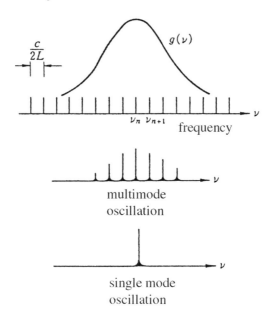

Figure 3.4. Laser spectra for different distributions of axial modes.

frequency. In addition to reflection losses at the mirrors, loss in the resonator occurs due to other loss mechanisms such as refraction and scattering. For this reason, the second expression in equation (3.7) is usually used.

In order to further quantify the phenomena in a laser oscillator, the situation of the oscillator can be treated as an electromagnetic wave in a resonant cavity, and analysed using a wave equation with the conditions at the cavity mirrors as the boundary conditions. The resonant electromagnetic fields, called *modes*, are comprised of sums of the fundamental field distributions in the cavity. In a cavity such as a Fabry–Perot resonator, which consists of two parallel mirrors separated by open space, components of the field that propagate at large angles to the resonator axis are lost because the sides of the resonator are open. Waves that travel in the direction of the axis, however, are bounded by the resonator mirrors. These modes are called *axial modes*, or *longitudinal modes*. The resonant frequencies of these modes are given by

$$\nu_n = \frac{cn}{2L} \quad (3.8)$$

where n is a positive integer and L is the cavity length. For a cavity with $L = 1$ m, the fundamental frequency is 150 MHz, and higher-order modes are separated by 150 MHz. Visible light frequencies are of the order of 10^{15} Hz, and so the resonant modes that exist in lasers have very large values of n. A gas laser has a relatively narrow gain band $g(\nu)$, but even in this case, the width is more than 1 GHz, caused by Doppler broadening. Thus, simultaneous oscillation on many

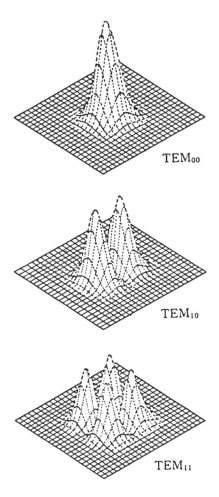

Figure 3.5. Calculated beam cross-sections for different laser modes, calculated using Hermite–Gauss mode expansion.

axial modes can occur, as shown in figure 3.4. For solid-state or liquid lasers, which have even broader $g(\nu)$, the number of oscillation modes can be of the order of thousands, or even tens of thousands.

Modes that are determined by the electromagnetic field distribution in the plane perpendicular to the resonator axis are called *transverse modes*. Unlike longitudinal modes, very high order transverse modes cannot be excited, and different transverse modes can be identified by their beam patterns. The lowest order mode, called the TEM_{00} *mode*, has a cross-sectional profile that is similar to a Gaussian distribution, and for this reason, the laser beam that comes from this mode is called a *Gaussian beam*. Higher-order modes, such as those shown in figure 3.5, have beam patterns with nodes whose number depends on the

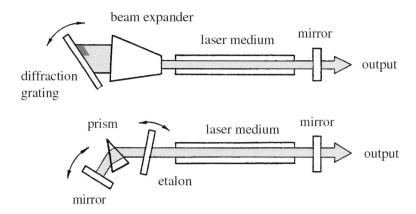

Figure 3.6. Different arrangements for reducing the spectral width of oscillation in a laser.

order of the mode. In addition, a beam with higher-order modes has increasingly large beam divergence. In most lasers, many high-order modes oscillate simultaneously, and so individual mode patterns cannot be identified. Only the increased beam divergence is observed.

Lasers have the highest coherence when they are operated in a single mode. Many commercially available lasers, however, cannot achieve single-mode oscillation. Generally, the effect of multi-longitudinal-mode oscillation is that the spectral width of the laser increases and the temporal coherence decreases. The effect of oscillation on multiple transverse modes is a decrease in spatial coherence, leading to increased beam divergence and degraded focusability.

3.1.2 Control of Laser Light

Control of the laser spectrum. A laser that operates in a single axial mode has a very narrow spectral width. However, in most lasers, there are a number of factors that contribute to spectral broadening. In Ar-ion lasers, for example, simultaneous oscillation of several different transitions at blue and green wavelengths can occur. In CO_2 lasers, oscillation on many vibrational and rotational transitions can be observed. In some liquid and solid-state lasers, the width of the gain band $g(\nu)$ can be from several tens of nanometres to over a hundred nanometres. These very wide-band lasers are called *tunable lasers*. In all of these lasers, the spectral output is controlled by inserting an optical band filter into the laser resonator. The oscillation wavelength can be varied by moving the transmission window of the filter.

In most lasers, the laser wavelength is controlled by using a device such as an etalon or a diffraction grating inside the resonator, as is shown in figure 3.6. Prisms can be used to control the laser wavelength, but their dispersion is small, and hence they can be used only for rough wavelength selection. Diffraction gratings have large dispersion, but together with large loss they have the additional disadvantage of being easily damaged by intense light. A better arrangement, shown in the figure, is to use a beam expander together with a diffraction grating. This reduces the possibility of damage to the grating surface, while keeping the advantage of high dispersion. An alternative arrangement is to use an etalon, which is a band filter that utilizes the interference that occurs between two parallel mirrors. By placing such a device in the cavity between the resonator mirrors, the oscillation frequency can be adjusted with resolution of less than 1 GHz.

For a laser oscillating in a *single* mode, the spectral width is determined ultimately by changes in the resonant frequency of the cavity that occur because of fluctuations in the length of the cavity. These fluctuations range from the extremely slow drift in cavity length produced by thermal expansion to fast fluctuations in the cavity length due to mechanical vibrations. In order to stabilize the frequency, a variety of thermal and mechanical stabilization techniques can be used. Stable oscillation with spectral width of less than 1 MHz requires, in addition, some form of feedback system in which frequency fluctuations are detected and fed back into the cavity.

Control of the pulse duration. The ability of lasers to generate light at a range of pulse widths, varying from continuous (cw) to femtosecond order, is another attractive feature of lasers. By compressing the duration of the laser pulse, it is possible to generate extremely large peak power.

The simplest method for producing pulsed oscillation is to use pulsed excitation. However, for the laser oscillation pulse to have the same width as the excitation pulse, the relaxation time of the laser medium must be much shorter than the excitation pulse, so that a quasi-steady-state equilibrium is achieved during excitation. The relaxation time for many solid-state laser media is of the order of 100 μs, and the output pulse shape differs from that of the excitation pulse in these lasers. By using a transient phenomenon, called 'spiking', pulses that are much shorter than the excitation pulse can be generated. In semiconductor lasers, 100 ps pulses can be obtained straightforwardly using current modulation. By using 'spiking', pulses with widths less than 1 ps can be generated. In excimer and nitrogen lasers, pulse widths of less than 10 ns can be achieved by using pulsed discharges that have fast rise times. In nitrogen lasers, 1 ns (or less) pulses can be generated by using travelling wave excitation.

In solid-state lasers, the laser medium has a relatively long relaxation time, and the technique called Q-switching is extremely effective for generating short, high power pulses. This technique, used in lasers such as Nd:YAG lasers, is shown in figure 3.7. The laser cavity contains a fast shutter, which is closed when excitation of the laser medium begins. During excitation, a large inverted population is generated in the laser medium because the closed shutter inhibits laser oscillation. The accumulated energy then is extracted in a short time by opening the shutter extremely quickly. In the case of flashlamp excited YAG lasers, the fast Pockels cell, an electro-optical switch, is used as the shutter, and

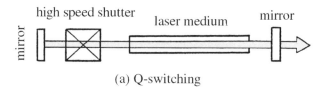

(a) Q-switching

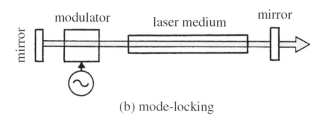

(b) mode-locking

Figure 3.7. Methods for generating short laser pulses.

pulse widths of less than 10 ns are achievable. For continuously excited YAG lasers, acousto-optical cells, which have very high repetition rates, can be used to produce pulse widths of 100 ns at pulse repetition rates of several kHz.

For producing pulses shorter than this, the most effective method is *mode-locking* [5]. Any laser operating on multiple axial modes has a collection of modes in the laser that are separated in frequency by $\Delta v = c/2L$, from equation (3.8). The presence of many modes at different frequencies results in beating between the modes, with the beat period $\Delta \tau$ being equal to the inverse of Δv, namely $\Delta \tau = 2L/c$. This period is the time that it takes for light to make one round-trip in the resonator. However, this beat is not observed usually because the phase of each mode fluctuates randomly. Mode-locking is the process in which the multiple axial modes are forced to be in phase at a particular time. This process produces a train of narrow pulses with a period of $\Delta \tau$.

The spectral width of a single mode-locked laser pulse is proportional to the number of locked axial modes. Hence, gas lasers, which have few axial modes, cannot be operated in very short pulses. Nd:YAG lasers, however, can be operated in pulses of the order of 50 ps. Tunable lasers, such as dye lasers or Ti:sapphire lasers, have very wide gain bandwidths, and can be used to generate ultrashort pulses, of 100 fs or less. For generation of pulses shorter than this, the method called *chirping* is used.

There are several types of mode-locking techniques. In one type, called *active mode-locking*, a modulator is placed inside the laser cavity, as shown in figure 3.7(b), and a frequency change of exactly $\Delta v = c/2L$ is applied. In another type, called *passive mode-locking*, a saturable absorber is inserted into the cavity. For lasers that are excited by another laser, such as dye lasers and

Ti:sapphire lasers, the excitation laser is mode-locked, and the cavity length is chosen to be an integer multiple of that of the pump laser. This is called *synchronous pumping*. In some of tunable solid-state lasers, such as Ti:Sapphire lasers, stable self-mode-locking can be achieved using a nonlinear effect in the laser material. Recently, lasers with very stable femtosecond pulses have become commercially available.

In the method called chirping, the chirped pulse is compressed by transmitting the radiation through a dispersive optical system, which contains elements such as diffraction gratings. The shortest laser pulse yet generated, with a width of 4 fs, was produced with this technique.

Control of beam quality. The beam quality of a laser generally is determined by the number of transverse modes. For the case of a laser oscillating in a single transverse mode, the directivity and focusability of the laser beam are determined by diffraction theory. However, for high-power lasers with a large active volume, higher-order modes always are present, and the beam divergence and focusability are poorer than the theoretical diffraction limit.

Because higher-order modes include components that propagate at large angles to the axis, the most effective way of suppressing higher-order modes is to insert an aperture into the resonant cavity in order to physically block these modes. However, doing this always reduces the output power. In order to achieve both good beam quality and high output power, it is necessary to use curved mirrors as the resonator mirrors. If the mirrors have radii of curvature r_1 and r_2, and the resonator length is L, the condition for obtaining a stable Gaussian beam is

$$0 \le \left(1 - \frac{L}{r_1}\right)\left(1 - \frac{L}{r_2}\right) \le 1. \tag{3.9}$$

In some lasers, such as solid-state lasers, the active laser medium shows a thermal lensing effect, and a correction to equation (3.9) is necessary.

However, by using the above method, it often is difficult to achieve TEM_{00} mode operation with high output power. In most cases, an oscillator–amplifier arrangement is the most effective method of achieving both these properties, especially for the case of high-power pulsed lasers. In this arrangement, a low-power oscillator is used to generate a beam with good quality that then is amplified further in a separate laser medium. A method of improving the beam quality is to place a spatial filter in the beam, with the aperture of the filter having a diameter that is close to the diffraction limited beam diameter. Instead of using an aperture, a mirror whose reflectivity varies in the radial direction also can be used effectively to control the beam quality. In addition, by operating the laser in a region which is unstable according to equation (3.9), and then using the radiation that is diffracted around a small resonator mirror as the output beam, high gain and large beam area lasers can be operated with good beam quality. This is called the *unstable resonator method*.

3.1.3 Gas Lasers

Gas lasers have several characteristic features: they can be used to generate laser radiation over a wide wavelength range, from X-rays to microwaves; their coherence is generally good, because the spectral line width is narrow; cooling and exchange of the laser medium is possible by circulating the gas; and electrical energy can be injected directly into the laser medium, by using discharge excitation. The features particular to each different type of gas laser are described below.

Neutral atom lasers. There are relatively few lasers based on transitions in neutral atoms. The only commercially available types are the He–Ne laser and the Cu-vapour laser.

He–Ne lasers can be operated in a stable cw mode, have high coherence and are inexpensive. However, their output power is limited to about 50 mW, and He–Ne lasers are being replaced by semiconductor lasers in most applications. Oscillation usually occurs on the red 632.8 nm line of the neon atom, but operation in the infrared, at 1.15 μm, 1.50 μm and 3.1 μm, and in the visible, at 612 nm (red), 594 nm (yellow) and 544 nm (green), also is possible. The laser normally is operated in a single transverse mode, but not in a single longitudinal mode. A notable exception is a single longitudinal mode, frequency-stabilized laser with spectral width of less than 10 kHz that was developed for specialized uses such as measurement.

Table 3.1. Wavelengths and typical output powers for commercially available Ar-ion and Kr-ion lasers.

Argon-ion laser		Krypton-ion laser	
Oscillation wavelength (nm)	Typical output power (W)	Oscillation wavelength (nm)	Typical output power (W)
Visible multiline	10.0	Infrared multiline	1.0
528.7	0.8	793.1–799.3	0.2
514.5	4.5	752.5	0.8
501.7	0.8	676.4	0.6
496.5	1.2	647.1	2.00
488.0	3.5	568.2	0.80
476.5	1.2	530.9	1.00
472.7	0.4	520.8	0.45
465.8	0.2	482.5	0.25
457.9	0.6	476.2	0.25
454.5	0.2	468.0	0.4
UV(333–363)	2.0	413.1	1.2
UV(275–305)	0.1	UV(337–356)	1.5

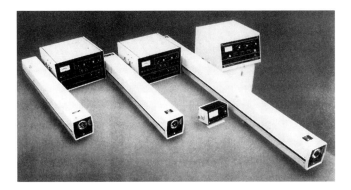

Figure 3.8. Examples of commercially available Ar-ion lasers (Coherent).

Cu-vapour lasers are pulsed lasers that oscillate on either a 511 nm green line or a 578 nm yellow line. The laser can be operated at a high repetition rate, up to 10 kHz, with output power of 200 W and laser efficiency of about 1%. The pulse width is in the range of 20–50 ns. Although this is the most powerful and efficient laser operating at visible wavelengths, it has to be operated at 1500°C, and the maintenance is relatively complicated. Hence, Cu-vapour lasers are not widely used, and their main area of application has been for test plants of uranium isotope separation. Output at 628 nm, though at reduced efficiency, is possible by changing the material in the laser from copper to gold.

Ion lasers. There are many kinds of lasers based on transitions in atomic ions, and laser action is possible at wavelengths distributed from the X-ray to the visible region. These lasers have the important advantage that short wavelengths can be generated. However, a large amount of energy is needed to generate the ions that produce the radiation and this poor efficiency is a significant disadvantage.

The most practical ion lasers are the Ar- and Kr-ion lasers. Typical values of the power and oscillation wavelengths for commercial lasers are shown in table 3.1. For Ar-ion lasers, the 488 nm and 514 nm lines are the strongest, and power of up to 25 W is achievable. In addition to this, output in the UV at up to 4 W also is possible. Figure 3.8 shows a photograph of a commercially available Ar-ion laser. This type of laser uses an electrical discharge to generate the ions, and has a small diameter discharge tube, and a large discharge current of 20–50 A. The combination of small tube diameter and large discharge current mean that a great deal of water-cooling is required, and the choice of material for the discharge tube also is important. The laser is operated with the discharge tube sealed, but small amounts of argon have to be added periodically because the gas pressure gradually decreases over time. The resonator mirrors are placed outside the discharge, and radiation from the discharge passes along the laser axis through Brewster windows mounted on the ends of the discharge tube. A Littrow

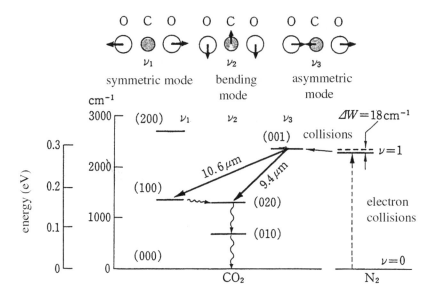

Figure 3.9. Vibrational modes and energy levels of the CO_2 molecule.

prism is inserted into the resonant cavity when single wavelength operation is desired. Furthermore, for single transverse mode operation, an etalon must be placed in the cavity. In addition to large high-power lasers, small air-cooled Ar-ions with powers of up to 100 mW are available.

Ion lasers based on metallic ions, such as Cd^+, Se^+, Zn^+ and Pb^+, also have been developed. In particular, the He–Cd laser is a practical laser, and can be used to produce low output power at the relatively short wavelengths of 442 nm and 325 nm. In addition to the basic He–Cd laser, there is a version in which laser action is obtained simultaneously on blue and red lines, balanced in a way so that the output light is 'white'.

Infrared molecular lasers. CO_2 lasers are the most well known example of the class of infrared lasers that use vibrational and rotational transitions of molecules as lasing transitions. The wavelengths available from these types of lasers range from the near infrared to the microwave region.

The CO_2 laser is the most efficient high-power gas laser, with efficiency of greater than 10% and maximum output power of 30 kW. Transitions between the three vibrational modes, v_1, v_2, and v_3 of the CO_2 molecule are used, as shown in figure 3.9. Oscillation can be achieved on the 10.6 μm and 9.4 μm bands, but it should be noted that these bands actually consist of many separate lines, because the vibrational levels shown in figure 3.9 are composed of many closely spaced rotational sub-levels. The transitions are divided using selection rules into the so-called P, Q and R branches, and the strongest oscillation lines are in the 10.6 μm P band, at the P(20) and P(22) transitions. Oscillation on a

Figure 3.10. An example of a commercially available TEA-CO_2 laser (Lumonics).

single line can be achieved by using a diffraction grating in the resonator cavity. The efficiency of exciting the CO_2 (001) mode can be improved by mixing the CO_2 gas with N_2 gas.

Both DC and RF discharges can be used for excitation in CO_2 lasers. Sealed-off operation is possible in the 20 W class lasers that are used for laser surgery. By using narrow discharge tubes, with diameter about 1–2 mm, the laser gain is increased, and the resonator mode in the cavity becomes a waveguide mode. This type of laser is called a waveguide laser.

For very large, high-power lasers, a transverse discharge tube is used. In this design, the discharge is generated between two long, parallel electrodes, and the gas circulates transversely. Therefore, the direction of the laser axis, the current direction and the gas flow direction are perpendicular to each other. This configuration is the most effective arrangement for a high-power laser that currently exists. In this transverse discharge tube, the pressure is very high, and the discharge gap between the electrodes is small, enabling a very uniform discharge to be generated. This is very suitable for generation of high power. Pulsed CO_2 lasers with this design can produce output beams with energy of more than 100 J, operating at pressures of up to 1 atm. This design is called a *TEA* (transversely excited atmospheric) *laser*. An photograph of a TEA-CO_2 laser is shown in figure 3.10.

A list of infrared molecular lasers is shown in table 3.2. The group of chemical lasers listed in the table use chemical reactions to produce population inversion. Although output power of up to 1 MW is available, these lasers are not used widely in industry. The molecular lasers in the submillimetre wavelength range can be based on either discharge excitation or optical excitation via absorption of CO_2 laser radiation at a particular wavelength. Using the latter excitation method, both pulsed and cw laser operation are possible.

UV molecular lasers. All of the different kinds of molecular lasers that

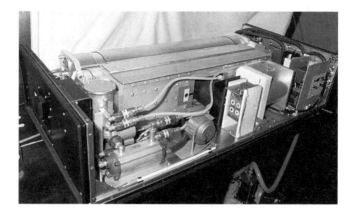

Figure 3.11. Interior of a discharge excited excimer laser (Lambda-Physik).

generate light at ultraviolet wavelengths use *electronic* transitions in molecules as the basis of the laser. *Excimer lasers* are based on compounds of rare gas atoms that form only in excited states. Because the ground state of excimer molecules dissociate rapidly, the density in the ground state is always zero, which makes excimer molecules suitable for use as the lasing species. Lasers using rare-gas halide excimers, shown in table 3.3, provide pulsed light at several different UV wavelengths with high power and reasonable efficiency. Lasers using KrF (λ = 249 nm) and XeCl (λ = 308 nm) produce average powers of 200–300 W at an efficiency of 3–4%. The typical pulse duration of these lasers is 10–30 ns.

A photograph of a commercially available laser is shown in figure 3.11. Rare-gas halide excimer lasers are operated with small amounts of halogen donor gases, such as HCl or F_2, mixed with rare gas at high pressures of up to

Table 3.2. Typical molecular lasers with their infrared wavelengths.

Laser gas	Type of excitation	Main oscillation wavelength range (μm)
CO_2	discharge	~ 10.6, ~ 9.6
CO	discharge	~ 5
N_2O	discharge	10.8
HF	chemical	2.4–3.2
HCl	chemical	~ 4
HBr	chemical	~ 4.3
H_2O	discharge, optical	27, 78, 118, etc.
DCN	discharge, optical	310, 336, etc.
HCN	discharge, optical	190, 195, etc.

Table 3.3. Typical characteristics of commercially available medium-size excimer lasers (Lumonics).

Laser	λ (nm)	Gas mixture	Pulse energy (mJ)	Average power (W)
F_2	157	F_2/He	12	1
ArF	193	F_2/Ar/He	150	30
KrCl	222	HCl/Kr/He	25	2.5
KrF	249	F_2/Kr/He	250	50
XeCl	308	HCl/Xe/He	200	40
N_2	337	N_2	7	0.4
XeF	351	F_2/Xe/He	125	25

Table 3.4. Materials used for solid-state lasers and their lasing wavelengths.

Active ion	Parent material	Oscillation wavelength (μm)
Nd^{3+}	YAG, YLF, glass GGG, GSGG, YVO	~ 1.06, ~ 1.3
Cr^{3+}	Al_2O_3 (ruby)	0.694
Er^{3+}	YAG, YLF, glass	~ 2.94
Ho^{3+}	YAG, YLF, GSGG	~ 2.1
Tm^{3+}	YAG	~ 2.1

several atmospheres. A transversely excited discharge tube is used. Pre-ionization discharges, generated by an array of small pins on each side of the main electrodes, are used to help produce a uniform discharge. The F_2 and N_2 lasers shown in table 3.3 use transitions in ordinary molecules rather than in excimers. The F_2 laser, with wavelength of 157 nm, is the shortest wavelength laser that is commercially available, while the N_2 laser, with wavelength of 337 nm, is a stable, easy-to-use laser source.

3.1.4 Solid-State and Semiconductor Diode Lasers

Solid-state lasers. Compared with gas lasers, solid-state lasers have several advantages. Firstly, the laser medium is made of condensed matter, and so high output power can be obtained from a small active medium. Also, the apparatus is simple, reliable and easy to maintain because optical pumping is used for excitation. In addition, the long lifetime of the upper level of the laser transition means that Q-switching can be used effectively to obtain high peak power. One disadvantage is that the wavelengths of solid-state lasers are limited mainly to the infrared. Table 3.4 shows the main solid-state laser materials together with

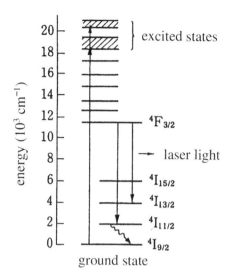

Figure 3.12. Energy level diagram of the Nd^{3+}:YAG crystal.

their corresponding lasing wavelengths. Although many new materials have been developed recently for use as the active medium, the Nd:YAG laser is still the most practical and well developed solid-state laser.

Figure 3.12 shows the energy level diagram for the Nd^{3+}:YAG (yttrium aluminium garnet: $Y_3Al_5O_{12}$) crystal. This four-level system is used for lasing, and excitation is by light from Xe or Kr lamp. The main laser wavelength is at $\lambda = 1.06$ μm, but it also is possible to obtain laser action at $\lambda \sim 1.3$ μm. Both cw and pulsed operation are possible. The general arrangement for the laser is shown in figure 3.13. The laser rod and the lamp are placed inside an elliptically shaped cylindrical mirror, at the focal points of the ellipse. For the case of cw excitation, power of up to 2 kW can be obtained in a single TEM_{00} mode, and 10 kW is possible in multi-mode operation. Using cw excitation together with an acousto-optic Q-switch, 200–300 mJ of pulsed power at kHz repetition rate is achievable. When flashlamp excitation and an electro-optic Q-switch are used, output of up to 3 J with pulse width of 5–7 ns is possible, although the pulse repetition rate in this case is limited to 20–50 Hz. This type of Nd:YAG laser is used mostly for research purposes. In addition, nonlinear crystals can be used to convert the fundamental radiation at 1.06 μm to the second-harmonic wavelength (532 nm), the third-harmonic wavelength (355 nm) and the fourth-harmonic wavelength (266 nm) with conversion efficiency of 40–50%.

The ruby laser ($\lambda = 694$ nm) uses a larger crystal for the lasing medium, and higher output energy of up to 20 J per pulse can be generated. However, because laser action occurs in a three-level system, the oscillation threshold is high and both cw and high repetition rate operation are difficult. The Nd:glass laser ($\lambda = 1.06$ mm) can generate even higher energy pulses, but for the same

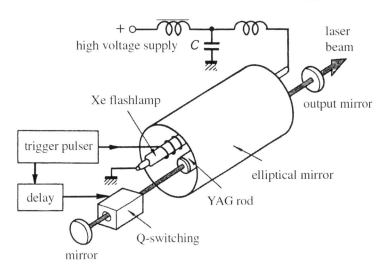

Figure 3.13. Physical arrangement of a flashlamp-excited, Q-switched Nd:YAG laser.

reasons as for the ruby laser, cw and high repetition rate operation are difficult. Other solid-state lasers include the Er:YAG laser ($\lambda = 2.94$ μm) and the Ho:YAG laser ($\lambda = 2.09$ μm), which are called 'eye-safe' lasers due to their long wavelength, but both of these lasers have inferior laser properties compared with the Nd:YAG laser.

Semiconductor diode lasers. Semiconductor diode lasers, also known as laser diodes (LD), are characterized by the following features: they can be operated by small current and have high efficiency; they are compact, easy to use and reliable; fast current modulation is possible; and they can be produced in large numbers at low cost. These lasers are an essential hardware component in the field of opto-electronics, and they are gradually replacing more conventional laser sources in many applications.

An example of a semiconductor diode laser is shown in figure 3.14. Basically, they consist of a GaAs diode with a pn junction. When current flows in the forward direction, emission occurs due to carrier recombination. In practical devices, a double-hetero junction (longitudinally) and a current narrowing structure (transversely) are used to confine the carriers and photons in a waveguide with a small cross-section. This structure also helps to lower the lasing threshold. In a TEM_{00} mode 10–100 mW of cw output can be produced.

Figure 3.15 shows different kinds of diode lasers with their wavelength ranges. The laser wavelength is determined by the gap width of the forbidden band. For example, in the case of the $Ga_{1-x}Al_xAs$ diode, the oscillation wavelength can be varied in the range of 680–900 nm by changing the composition ratio x. The gain bandwidth is about 10 nm, and the diode laser

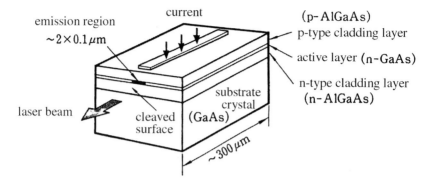

Figure 3.14. Schematic diagram of a GaAs diode laser with a double-hetero junction.

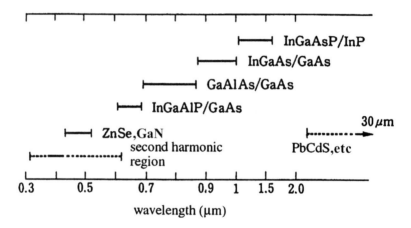

Figure 3.15. Oscillation wavelength ranges for the main types of semiconductor diode lasers.

usually oscillates in a few modes. Because the oscillation wavelength also varies with diode temperature and current, stable single-mode operation can be obtained only by using a *distributed feedback* (DFB) scheme, in which a grating is placed in the waveguide. GaAs-based lasers, with wavelengths between 630 nm and 1.5 μm, are used for opto-electronics and have reasonably good laser properties. Diode lasers based on the other materials shown in figure 3.15 generally have inferior properties. Pb–chalcogenide lasers can produce wavelengths between 2.8 μm and 30 μm, and are used for spectroscopic purposes, but liquid nitrogen cooling is essential. Recently, ZnSe and GaN lasers have been demonstrated to work at room temperature, generating blue and green wavelengths, and in the near future these lasers will be commercially available.

The upper limit of power for a single diode laser operating in a TEM$_{00}$ mode is about 200–300 mW, because high power can damage the diode itself. However, output power of several watts can be achieved by a broad-area stripe scheme. Furthermore, a 100 W class laser can be obtained by using 1D and 2D arrays. However, these high-power sources have relatively poor coherence, and their dominant application is pumping of Nd:YAG lasers.

Highly efficient pumping of Nd:YAG is possible using laser diodes because this crystal has an absorption band at about 800 nm. This efficient pumping means that compact, easy-to-handle Nd:YAG lasers can be produced. For high-power systems, efficiencies of more than 20% have been obtained. In addition, second-harmonic wave generation can be performed with high efficiency because oscillation in the TEM$_{00}$ mode can be readily achieved using *end pumping* of the laser crystal. By using this scheme, compact lasers with output powers of more than 100 mW have been produced. *Side pumping* using an array of diode lasers has been used to produce cw output of more than 10 kW. A 6 W cw green laser has already been produced commercially, and this laser source soon might replace the Ar-ion laser. Q-switched Nd:YAG lasers that use diode laser excitation can produce laser output that is comparable with commercially produced flashlamp pumped lasers. Such lasers, now at the research stage level, have produced output energies of 1 J at repetition rates of tens of Hz.

Diode-pumped solid-state lasers are widely viewed as the next generation high-power lasers. At present, large-scale devices are expensive because of the high cost of the diodes themselves. The cost of these diodes is expected to fall, however, when they are mass-produced. There is rapid progress in this field, and compact, reliable, easy-to-use lasers should become available in the near future.

3.1.5 Tunable Lasers

As discussed previously, all laser media have gain $g(\nu)$ with a finite spectral width, but truly tunable output only can be achieved when the gain width is particularly large. In this section, we will discuss dye lasers and the relatively new tunable solid-state lasers. Although not discussed in detail here, the semiconductor diode laser also can be considered a type of tunable laser, because the resonant wavelength can be changed continuously by altering the semiconductor composition, diode current and diode temperature.

The basic arrangement for a tunable laser is shown in figure 3.6. A laser medium with a wide gain bandwidth $g(\nu)$ is placed in a resonant cavity containing a wavelength-selective element. This arrangement allows continuously tunable laser output to be generated. Tunable lasers such as these are a necessary piece of hardware for many of the laser spectroscopic measurement systems described later in this book [6].

Dye Lasers. Dye lasers are the only practical laser sources that use a liquid as the laser medium [7]. Excitation of the dye molecules is achieved optically,

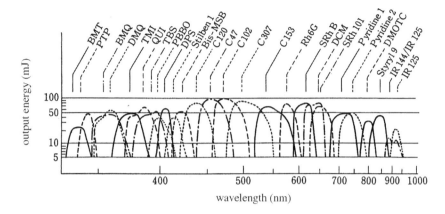

Figure 3.16. Range of available wavelengths for an excimer laser-pumped dye laser (graph courtesy of Lambda-Physik).

using flashlamps or laser sources such as excimer lasers and Nd:YAG lasers. The organic molecules that are used for the dye have complicated structures, and the combination of the vibrational and rotational structure of the dye molecules results in a wide, continuous fluorescence spectrum. The laser wavelength can be varied continuously within this wavelength region. The tunable range of a single laser dye is of the order of tens of nanometres, and laser radiation from 320–1000 nm can be generated using a series of laser dyes. Figure 3.16 shows the types of dyes that are used to cover this wavelength range. Excitation of the laser dye is provided by one of three main light sources; Ar- or Kr-ion lasers, which produce cw radiation at visible and UV wavelengths; short pulsed high-power lasers with pulse widths of the order of ns; and flashlamps.

Continuous-wave dye lasers are an essential tool for high-resolution spectroscopy, because they can be operated in stable, single-longitudinal, single-transverse modes. A ring oscillator, such as that shown in figure 3.17, can be used for stable single-mode operation, and spectral width of 1 MHz or less can be achieved using a feedback system for frequency stabilization. Using mode-locking, the same oscillator arrangement can be used to achieve extremely short pulse widths of less than 100 fs.

Lasers such as excimer lasers, Q-switched Nd:YAG lasers and N_2 lasers are used as pumping sources for short pulse-width, high repetition-rate dye lasers. Because high gain can be readily achieved in many dyes, it is possible to generate laser radiation over the wide wavelength range shown in figure 3.16. Also, because the instantaneous power is extremely large, nonlinear frequency conversion can be used to widen the wavelength range even further.

Dye lasers that use flashlamp pumping are characterized by comparatively long pulse widths (1–10 μs) and large pulse energies of the order of 1 J.

Figure 3.17. An example of a commercially available Ar-ion laser pumped ring dye laser (Coherent).

However, thermal fluctuations in the dye lead to poor coherence, and high repetition rates cannot be obtained. In general, because the laser medium is a liquid, changing of the laser medium is comparatively easy compared with solid-state lasers. In practice, however, the dye inevitably degrades during use, and so has to be changed frequently.

Tunable solid-state lasers. The colour-centre laser is a well-known example of a tunable solid-state laser, but although it has been commercially available for many years, it suffers from the need for low-temperature operation and instability of the laser medium. Recently, attention has focused on the newly developed solid-state laser sources listed in table 3.5, which have stable laser media together with wavelength tuning ranges and efficiencies similar to those of dye lasers. Because these laser devices have the common feature that the ground level is broadened due to coupling with lattice vibrations in the crystal (phonons), they often are called *phonon-terminated lasers* [8].

Table 3.5. Materials used for tunable solid-state lasers and their tuning wavelength ranges.

Material	Tunable wavelength range (nm)
Alexandrite (Cr^{3+} : $BeAl_2O_3$)	680–800
Cr : LiCAF (Cr^{3+} : $LiCaAlF_6$)	750–900
Cr : LiSAF (Cr^{3+} : $LiSrAlF_6$)	700–850
Ti : sapphire (Ti^{3+} : Al_2O_3)	650–1050
Forsterite (Cr^{4+} : Mg_2SiO_4)	1100–1300
Co^{3+} : MgF_2	1500–2300

The first laser source to be developed amongst this group was the *alexandrite laser*, which uses a Cr-doped laser crystal that is similar to ruby. However, unlike ruby, the crystal has a four-level system including the phonon band, so that both cw and high repetition-rate operation are possible. By using flashlamp pumping and Q-switching, output energies of 0.5–1 J at a repetition rate of 20 Hz can be achieved. The more recently developed Cr:LiSAF and Cr:LiCAF crystals have the additional advantage that they possess absorption bands close to 650 nm, and so laser diode excitation, in addition to flashlamp excitation, is possible.

Amongst the new types of laser crystals, most attention has been focused on Ti:sapphire. Because it has high gain and possesses an absorption peak at green wavelengths, both Ar-ion lasers and frequency doubled Nd:YAG lasers can be used to provide efficient excitation. Excitation by an Ar-ion laser has been used to generate tunable radiation over the wide range of 650–1050 nm. In addition, single-mode operation is far more stable than for the dye laser. When a Q-switched YAG laser is used as the excitation source, pulse power of tens of megawatts in nanosecond order pulses is achievable. Perhaps the most important contribution of Ti:sapphire lasers to the field of laser engineering is ultrashort pulse generation. Ti:sapphire laser systems with extremely high-power (>10 TW) femtosecond pulses have been developed.

3.2 Nonlinear Wavelength Conversion Devices

Nonlinear optics is a field that really began with the observation of nonlinear effects in materials produced by the high intensity and high coherence of laser light. It since has become one of the most important fields in laser science. In this section, an overview of nonlinear optical effects will be given, concentrating on those used for wavelength conversion devices [2–4].

3.2.1 Nonlinear Optical Effects

Nonlinear effects in ferromagnetic materials have been used in electronics for many years, but the magnetic dipole moment does not respond at optical frequencies. In this frequency region, the nonlinear dependence of the electrical polarization of a dielectric material on an applied electric field is used.

The polarization P in a dielectric material due to an electric field E is given by

$$P = \varepsilon_0 (\chi^{(1)} \cdot E + \chi^{(2)} : EE + \chi^{(3)} \vdots EEE + \cdots). \tag{3.10}$$

In isotropic materials, the electric susceptibility χ is scalar, and its even orders are known to be zero. Nonlinear phenomena due to $\chi^{(2)}$ are observed only in crystals that have a lack of inversion symmetry. In these cases, $\chi^{(2)}$ is a rank 3 tensor, and the nonlinear part of equation (3.10) can be written as

$$P_{NLi} = \varepsilon_0 \sum_j \sum_k \chi_{ijk}^{(2)} E_j E_k. \tag{3.11}$$

Because $\chi^{(2)}$ is a tensor, the relationship between P and E is very complex. For simplicity, though, it can be assumed to be a scalar relationship and in this case, an electric field $E = E_0\cos\omega t$ will produce polarization in the dielectric medium given by

$$P_{NL} = \varepsilon_0 \chi^{(2)} E_0^2 \cos^2(\omega t) = \frac{1}{2}\varepsilon_0 \chi^{(2)} E_0^2 (1 + \cos 2\omega t). \quad (3.12)$$

In other words, a polarization (electric dipole moment) with frequency 2ω exists in the medium. This polarization is the source of the second harmonic wave generated in the crystal. Also, if two electric fields with different frequencies, ω_1 and ω_2, are combined in the dielectric medium, a sum-frequency wave, with frequency $\omega_1 + \omega_2$ and difference frequency wave, with frequency $\omega_1 - \omega_2$, will be generated.

When an electric field E_0 is introduced into a crystal of thickness L, the power of the second harmonic wave $I_{2\omega}$ can be found by solving a nonlinear wave equation and is given by

$$I_{2\omega} = K \left|\chi^{(2)}\right|^2 (E_\omega)^4 \frac{\sin^2 (\Delta k L / 2)}{(\Delta k / 2)^2}. \quad (3.13)$$

In this expression, $\Delta k = k_2 - 2k_1$, where k_1 and k_2 are the wave numbers of the fundamental and second harmonic waves. If the refractive index of the crystal at the two wave frequencies is n_1 and n_2, then $k_1 = 2\omega n_1/c$ and $k_2 = 2\omega n_2/c$. For the case of $\Delta k = 0$, or

$$n_1 = n_2 \quad (\text{or} \quad 2k - k_2 = 0) \quad (3.14)$$

the $\sin^2 (\Delta kL/2)/(\Delta k/2)^2$ term becomes equal to L^2, and so the intensity of the second harmonic wave $I_{2\omega}$ increases proportionally with the square of L. $I_{2\omega}$ cannot become large unless this condition is satisfied. Equation (3.14), which must be satisfied for efficient second-harmonic generation, is called the *phase (or index) matching condition*. In physical terms, this is the condition that the 2ω wave polarization induced in the crystal by the ω wave contributes in phase to generation of the 2ω wave. The phase matching condition for optical mixing of two waves at frequencies ω_1 and ω_2, is $k_3 = k_1 + k_2$, where k_1, k_2 and k_3 are the wavevectors for the respective waves.

A similar nonlinear effect called the optical parametric effect can be described in a similar way. For this discussion, it is assumed that three plane waves, with fields E_s, E_i and E_p, propagate in a nonlinear medium in the z direction, with the angular frequencies of the waves satisfying the condition

$$\omega_s + \omega_i = \omega_p. \quad (3.15)$$

In this situation, the ω_p wave is generated by the interaction of the ω_i and ω_s waves with the nonlinear susceptibility $\chi^{(2)}$. The interaction between these waves is called a *parametric effect*. This is the most general effect observed in nonlinear media. If the phase matching condition $k_p = k_s + k_i$ is satisfied, this effect will be significant. The three waves are usually called the *pump* (ω_p), the

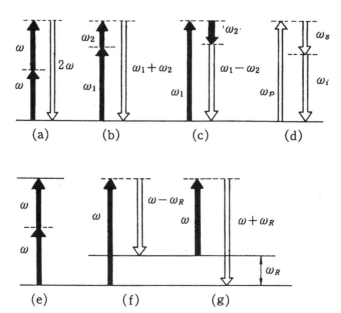

Figure 3.18. Different types of multi-photon processes. In each diagram, the black arrows and white arrow indicate input and output light respectively. The different schemes are: (*a*) second-harmonic generation, (*b*) sum-frequency mixing, (*c*) difference-frequency mixing, (*d*) parametric processes, (*e*) two-photon absorption, (*f*) stimulated Raman scattering (Stokes line), and (*g*) stimulated Raman scattering (anti-Stokes line).

signal (ω_s) and the *idler* (ω_i). The following relationships can be derived using a nonlinear wave equation.

$$\left. \begin{aligned} \frac{dE_s}{dz} &= -iK_s E_p E_i{}^* \\ \frac{dE_i}{dz} &= +iK_i E_s E_p{}^* \\ \frac{dE_p}{dz} &= -iK_p E_i E_s \end{aligned} \right\}. \tag{3.16}$$

In this expression, * indicates the complex conjugate and

$$K_n = \frac{\omega_n}{2}\sqrt{\frac{\mu_0}{\varepsilon_n}}\,\varepsilon_0\chi^{(2)} \tag{3.17}$$

where $n = s, i, p$. The amplification of the signal and idler waves can be derived by writing the electric fields as $A_n = (n_n/\omega_n)^{1/2} E_n$, with the amplitudes at $z = 0$ being given as $A_s(0)$ and $A_i(0)=0$. For the case when the amplitude of the pump wave does not depend on z,

$$A_s = A_s(0)\cosh(\gamma z/2)$$
$$A_i^* = A_s(0)\cosh(\gamma z/2)$$
(3.18)

where

$$\gamma = \chi^{(2)} E_p(0)(\varepsilon_0\mu_0\omega_s\omega_i/n_s n_i)^{1/2}.$$
(3.19)

This indicates that both the signal wave and the idler wave will be amplified exponentially if $\gamma z \gg 1$. The factor γ is the power gain, as can be seen from squaring equation (3.18).

If the frequency condition in equation (3.15) and the phase–matching condition in equation (3.14) are both satisfied inside the nonlinear crystal, amplification at ω_s and ω_i will occur. This is called *optical parametric amplification*. Coherent radiation can be generated at these frequencies by placing the medium inside a resonant cavity. This light source is called an *optical parametric oscillator* (OPO), and the oscillation wavelength can be tuned over the range given by equation (3.15).

The above discussion is based on classical electromagnetic theory. These nonlinear optical effects also can be explained in quantum mechanical terms, using the kinds of *multi-photon processes* shown in figure 3.18. By considering the energy exchange process in the medium, even the stimulated Raman scattering processes shown in (f) and (g), can be thought of as parametric effects. A coherent light source called a *Raman laser* has been developed, based on these principles.

3.2.2 Higher Harmonic Generation and Frequency Mixing

Table 3.6 shows the properties of typical nonlinear optical crystals [9]. As can be understood from equation (3.13), desirable properties for efficient *second-harmonic generation* (SHG) include large nonlinear susceptibility, high

Table 3.6. Comparison of characteristics of nonlinear optical crystals.

Crystal	Transmission range (μm)	Nonlinear coefficient*	Threshold for optical damage** (GW cm^{-2})
KDP	0.2–1.5	1	0.4
LiNbO$_3$	0.4–5.0	10	0.01–0.04
BBO	0.19–3.0	4.4	3–5
LAP	0.2–1.9	1.9	10–15
LBO	0.16–2.6	2.6–2.9	5–10
LiIO$_3$	0.3–5.5	12	0.01–0.05
KTP	0.35–4.5	11–15	0.5–1.0

* relative to that of KDP.
** for a 10 ns YAG laser pulse.

transmission at the fundamental and second-harmonic frequencies, satisfaction of the phase matching condition, and high damage threshold.

Usually the phase matching condition can be satisfied in a birefringent crystal if the crystal has the property that the ordinary wave and the extraordinary wave have the same refractive index. For example, in single-axis crystals such as KDP and BBO, the electric field of the fundamental beam is directed in the ordinary direction (i.e. polarized perpendicularly to the incident light plane), while the SHG wave has its electric field in the extraordinary direction (i.e. polarized parallel to the incident light plane). Hence, the phase matching condition is satisfied when the light enters the crystal at a particular angle. It is possible to calculate the phase matching angle from the dispersion of the crystal, but we will leave such a calculation to more specialized textbooks.

As can be seen from equation (3.13), the efficiency of second-harmonic generation is proportional to the square of both the incident laser power I_ω and the nonlinear susceptibility $\chi^{(2)}$. Although it is possible to increase I_ω by focusing the beam with a lens, this requires good laser beam quality, and the incident angle must be within the allowance angle for the phase matching condition. For cw lasers, I_ω is small, and so an intra-cavity configuration is required to increase its magnitude. The upper limit to I_ω is determined by the damage threshold of the crystal.

Amongst the crystals shown in table 3.6, BBO and KTP have properties that make them especially suitable for second-harmonic wave generation. The advantages of BBO are that it is transparent down to ultraviolet wavelengths, and it has a high damage threshold. Phase matching for the second-harmonic generation is possible down to $\lambda \sim 205$ nm, and frequency mixing is possible down to $\lambda \sim 190$ nm. KDP has the advantages of a larger nonlinear susceptibility, and a larger range of angles that satisfy the phase matching condition, but it cannot be used for wavelengths shorter than about 400 nm. Its longer wavelength limit, however, is about 5 µm.

High efficiency can be achieved for generating higher harmonics of the Nd:YAG laser. For Q-switched pulsed versions of this laser, stable SHG output can be obtained with efficiency of up to 50%. For cw versions, efficiency of 50% can be obtained by placing the crystal inside the laser resonator. For tunable lasers, such as dye lasers and Ti:sapphire lasers, second-harmonic generation is used to extend the tunable range of the laser. Although SHG output can be obtained for cw tunable lasers, much more stable and efficient SHG output is achievable with the pulsed versions.

Higher harmonic generation and sum-frequency mixing can be used to generate coherent radiation at shorter wavelengths. The third harmonic of the Nd:YAG laser is produced by mixing the fundamental and second-harmonic beams together in a separate crystal. Difference-frequency mixing can be used to generate coherent radiation at UV wavelengths, where there are few good tunable light sources. Recently, tunable IR radiation has been generated at wavelengths up to 5 µm by difference-frequency mixing of the beams from two laser diodes in a KTP crystal. Although the output power yet achieved is very

Nonlinear Wavelength Conversion Devices

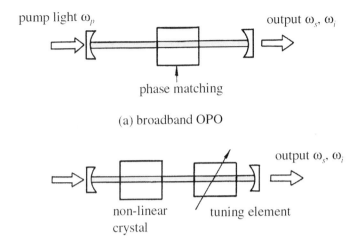

Figure 3.19. Schematic diagram of an optical parametric oscillator (OPO).

small, these are useful sources for IR absorption spectroscopy because their spectral purity is very high. The wavelength range can be extended up to 18 μm using this scheme with $AgGaSe_2$ crystals, but at poorer efficiency [10].

3.2.3 Optical Parametric Oscillators

Optical parametric oscillators (OPO) have a long history in research as coherent tunable light sources, but it is only relatively recently that they have been developed into commercial laser sources. The recent advances are due to the development of high-quality crystals such as KTP and BBO. In addition, because the development of narrowband OPO devices that can be used in spectroscopic measurements is continuing, these devices are considered to be the best prospects for the future tunable laser sources.

The basic arrangement for an OPO laser is shown in figure 3.19(a). Q-switched YAG lasers are the most suitable source for excitation. The phase matching condition determines the relationship between the frequencies of the signal wave ω_s and the idler wave ω_i. Figure 3.20 shows the relationships between the phase matching angle and the wavelengths of the signal and idle waves, for the case where the third (355 nm) and the fourth (266 nm) harmonic of Nd:YAG laser are used. In principle, an optical resonator can be made resonate at both ω_s and ω_i, or at only one of the frequencies. Most commonly single-frequency resonator is used because it is difficult to have reflect elements that are efficient at both frequencies, and also because the stability oscillation is poor for two frequency systems. Generation of the signal wave be achieved with conversion efficiency of more than 30% using pulsed Nd:Y laser excitation.

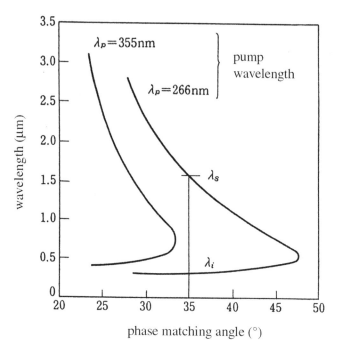

ure 3.20. The phase matching relationship for the signal and idler wavelengths, λ_s and
or an OPO processes in a BBO crystal.

or the arrangement shown in figure 3.19(a), the spectral width of the
s determined by the phase matching condition and hence can be fairly
order to obtain narrowband output, a tuning element has to be inserted
ivity, as shown in figure 3.19(b). Oscillation in a single longitudinal
ssible, but high gain is required for this. In order to achieve high gain
device, very high excitation light density is required but until
was difficult because the crystal itself was damaged by the
it. However, recent developments have made this possible, and
PO lasers now are available. Because the output from the
ill in these systems, further amplification or injection-locking
ied to increase the output power.

Scattering

attering and *stimulated Brillouin scattering* are types of
frequency of the scattered light is different from that of
These are shown in figure 3.18(f) and (g). In Brillouin
attered, and the frequency shift is of the order 1 cm^{-1}.
scattering is not particularly useful as a frequency
fts for Raman scattering are much larger, and this

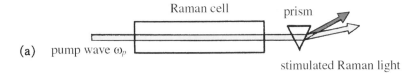
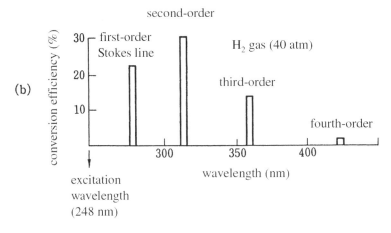

Figure 3.21. (*a*) Schematic diagram of a Raman shifter, and (*b*) Raman shifts for KrF excimer laser light in H_2 gas.

section will concentrate on frequency conversion devices based on Raman scattering.

In Raman shifting devices, the most commonly used conversion medium is high-pressure hydrogen gas. The vibrational Raman shift for hydrogen is large (4155 cm^{-1}) and absorption is small over a wide range of wavelengths. In addition, it is possible to utilize the rotational Raman shift (587 cm^{-1}) to access the infrared region. Other gases, such as deuterium (2991 cm^{-1}) and methane (2916 cm^{-1}) also are useful.

Figure 3.21(*a*) shows a practical Raman shifting device. The laser beam is passed through a high-pressure gas cell, which typically has a length of 0.5–2 m and a resonant cavity is not used. The frequency-shifted beam propagates in the same direction as the original beam. Because Raman cross-sections become larger in the ultraviolet, the beam from a UV laser such as a KrF excimer laser often is used. The output frequencies for this case are shown in figure 3.21(*b*). The longer wavelength components are called *Stokes lines*, and the shorter wavelength components are called *anti-Stokes lines*.

Although high absorption power, of the order of 1 MW, is required for practical Raman shifting, devices with high conversion efficiency can be constructed relatively simply. At shorter wavelengths, ultraviolet wavelengths as short as 120 nm can be obtained by using high-power excimer lasers and Raman

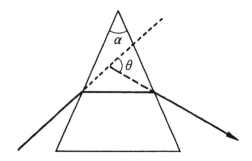

Figure 3.22. Refraction of light by a prism.

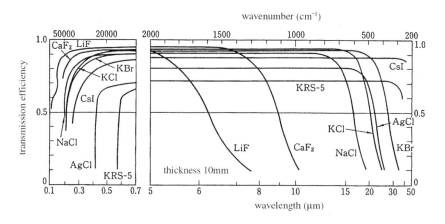

Figure 3.23. Transmission ranges for different optical crystals. Data taken from Rika-Nenpyo, Chronological Scientific Tables, ed. National Astronomical Observatory in Japan (Maruzen, 1986).

shifters. At infrared wavelengths, however, the Raman gain is much reduced and multipass cells that redirect the beam through the gas cell many times are used to increase the optical path length, and hence the conversion efficiency.

3.3 Optical Elements and Optical Instruments

3.3.1 Dispersion Elements and Spectrometers

Two types of optical elements are used to disperse light according to wavelength: *prisms* and *diffraction gratings*. The spectroscopic instruments that use these elements are called spectrometers [11].

Prisms. When a triangular prism, with angle α, is illuminated symmetrically, as shown in figure 3.22, the loss and the aberration of the transmitted light is minimized and the dispersion is maximized. In this arrangement, if the angle

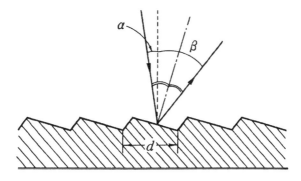

Figure 3.24. Diffraction of light at an echellete grating with blazed grooves.

between the incident and transmitted beams is θ, and the refractive index of the prism is $n(\lambda)$, a quantity called the *dispersivity* D_λ can be defined as

$$D_\lambda = \frac{\partial \theta}{\partial \lambda} = \frac{2\sin(\alpha/2)}{[1 - n^2 \sin^2(\alpha/2)]^{1/2}} \frac{\partial n}{\partial \lambda}. \tag{3.20}$$

The prism material should have large dispersion $\partial n/\partial \lambda$, and high transmission over a large range of wavelengths. Also, it is desirable that the material be optically homogeneous, and physically and chemically stable. Figure 3.23 shows the wavelength dependence of the transmission of typical prism materials. High transmission can be obtained for the wavelength range 250–2800 nm using natural crystals. For wavelengths less than about 120 nm, prisms cannot be used because no highly transmissive crystals exist for these wavelengths.

Diffraction gratings. Diffraction gratings are optical elements that have a surface consisting of a pattern of periodic grooves, and are used as reflection elements in spectrometers. Most gratings have a reflective surface with a saw-tooth pattern, as shown in figure 3.24. This kind of blazed grating has large diffraction efficiency in one particular direction, and is called an *echellete grating*.

If light is incident on the grating surface at an angle α, as shown in figure 3.24, the angle of the diffracted light β is given by

$$\sin \beta = -\sin\alpha + m\lambda/d. \tag{3.21}$$

In this expression, d is the periodicity (or pitch) of the grating surface, λ is the wavelength of the incident light and m is the diffraction order ($m = 0, \pm 1, \pm 2, \ldots$). For the case of $m = 0$ and $\beta = -\alpha$, there is no dispersion. The dispersion becomes progressively larger as the value of m increases. By differentiating equation (3.21), the dispersivity becomes

$$D_\lambda = \frac{\partial \beta}{\partial \lambda} = \frac{m}{d \cos\beta}. \tag{3.22}$$

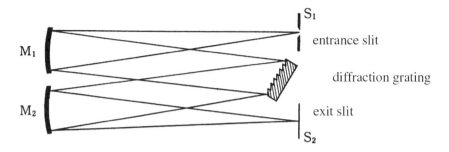

Figure 3.25. Physical arrangement of a Czerny–Turner monochromator.

Therefore, the dispersion is large for small values of d and large values of m. Compared with a prism, a diffraction grating has large dispersion, and the linearity in dispersion also is good. However, because the light is diffracted in many orders, the diffraction efficiency is poor. To overcome this problem, the grating surface is usually blazed so that the diffraction efficiency is large for a particular wavelength.

Echellete gratings for visible and ultraviolet wavelengths require surfaces with 1000–3000 grooves per millimetre, and extremely sophisticated machining technology is used to manufacture large area gratings. Recently, it has become possible to make holographic diffraction gratings by using laser interference techniques. Gratings with spherical surfaces are used for VUV spectrometers, as described below.

Spectrometers. A spectrometer is an instrument that is used to spectrally analyse light. Figure 3.25 shows an example of a *Czerny–Turner spectrometer*, which is representative of spectrometers that use diffracting gratings as the light dispersing element. Light that passes through an entrance slit is converted into a parallel beam by a mirror M_1, and directed onto a diffraction grating. The dispersed light is then focused on the surface S_2. When an exit slit is placed at S_2, the instrument can select one particular wavelength component of the incident light. If an image of the total spectrum is desired, photographic film or an imaging device such as a diode array or a CCD camera can be placed at S_2.

There are also several other types of spectrometers. Instruments used for very short wavelengths consist of just slits and a spherically shaped grating. The most well known example of this type of spectrometer is the *Rowland circle*, shown in figure 3.26. For a grating with radius of curvature R, the circle must have a radius of $R/2$. All the diffraction orders are focused to a single point on the surface of the circle. In this instrument, if photographic film is used to capture the entire spectrum, then the film too must be placed on this circle.

For the spectrometer shown in figure 3.25, the brightness, F, of the spectrometer is determined by the ratio of the focal length of the mirrors, f, and the effective size of the grating, D, giving $F = f/D$. The dispersion at the surface

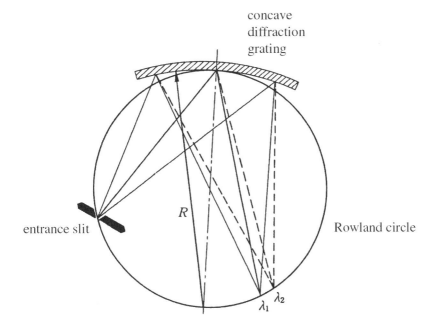

Figure 3.26. A Rowland circle.

S_2 is given by fD_λ. The ultimate resolution is determined by the product of the diffraction order of the grating m and the total number of grooves N on the diffraction grating itself.

3.3.2 Interferometers

Interferometers have been used as optical instruments for more than 100 years, but their performance has improved greatly since it became possible to use highly coherent lasers as light sources. Interferometers can be used to directly measure distance, displacement and vibration, and from these quantities density, temperature, refractive index and electron density can be inferred. They can also be used as spectroscopic elements. The three main types of interferometers are shown in figure 3.27. Most interferometers are a simple combination of planar mirrors, but each mirror must have a surface accuracy of 1/10 to 1/100 of the light wavelength. Similar accuracy is required for the mirror mountings. This high degree of accuracy is required because the interference is extremely sensitive to changes in the phase of the light.

The *Michelson interferometer*, shown in figure 3.27(a), has the most basic interferometric arrangement, and is famous for having been used in several renowned fundamental physics experiments. In this arrangement, light from a diverging light source is used, and interference fringes are formed at the surface S, at the focal point of the lens L. The interference fringes are a series of

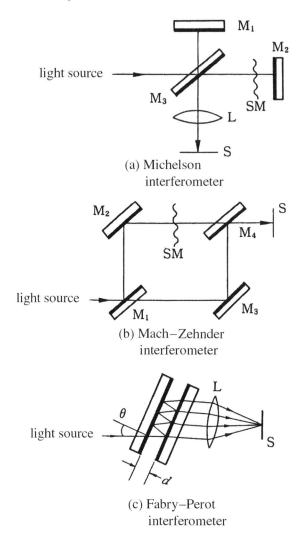

Figure 3.27. Schematic diagrams of the main types of interferometers (M: mirror, L: lens, S: screen, SM: sample to be measured).

concentric circles. When the mirror M_2 is moved by a distance of $\lambda/2$, the fringes at S are shifted by a single position (i.e. each fringe moves into the position of the neighbouring fringe). Consequently, by detecting the number of fringes that pass a certain position, the movement of the mirror can be measured. When the optical path difference for the two paths is large, the interference fringes are extremely narrow, and when the optical path difference becomes zero, the fringes disappear and the entire area becomes uniformly bright.

When a collimating lens is inserted into the beam paths, the instrument is called a *Twyman–Green interferometer*. In this instrument, a sample is placed in one beam path, in front of the mirror. Differences in the optical paths of light passing through each point of the sample leads to a set of contour-like interference patterns on the screen. These instruments are widely used to verify the quality of optical components, by checking the smoothness of the material surface, or the uniformity of the refractive index.

For a source with a spectrum $S(\nu)d\nu$, where ν is frequency, the intensity of light $I(x)$ observed at the focal point of the lens, as a function of the optical path difference x, is given by

$$I(x) = \int_0^x [1 + \cos(2\pi\nu x)] S(\nu) d\nu = \int_0^x S(\nu) d\nu + \int_0^x S(\nu)\cos(2\pi\nu x) d\nu. \quad (3.23)$$

The first term on the right-hand side of the above expression is simply half the intensity, $I(0)$, and so equation (3.23) can be written as

$$I(x) = \frac{1}{2} I(0) = \int_0^x S(\nu)\cos(2\pi\nu x) d\nu. \quad (3.24)$$

By using a Fourier transform, $S(\nu)$ can be written as

$$S(\nu) = 4\int_0^x [I(x) - \frac{1}{2} I(0)]\cos(2\pi\nu x) d\nu. \quad (3.25)$$

Hence, it can be seen from these equations that the spectrum of the incoming light can be determined by observing the intensity change $I(x)$ as the optical path difference x is varied. This is called *Fourier Spectroscopy* [6,12]. The advantage of this method is that the entire spectrum is detected as a single bright point, and so signals with good signal-to-noise ratios can be obtained. This is particularly useful at infrared wavelengths, where good detectors do not exist. Fourier transforms are performed easily using modern computers, and this measurement principle has been developed into a widely used measurement technique called *Fourier transform infrared (FT-IR) spectroscopy*.

Figure 3.27(b) shows a *Mach–Zender interferometer*. In this instrument, a sample, which may be very large, is placed between mirrors M_2 and M_4, and the change in refractive index due to the presence of the sample produces a two-dimensional distribution of interference fringes on the screen. This type of instrument is widely applied for measurement of large objects such as gas flows, plasmas, shock waves and combustion fields.

The arrangement shown in figure 3.27(c), in which the interference fringes are produced by multiple reflections of the incident light between two parallel mirrors, is called a *Fabry–Perot interferometer*, or more simply, an *etalon*. A related device is the solid etalon, which is a single piece of high-quality optical glass with high-reflection coatings on both sides of the glass. Light that enters the interferometer is transmitted through the interferometer after multiple

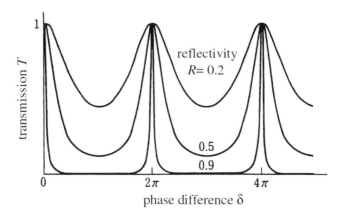

Figure 3.28. Transmission properties of a Fabry–Perot interferometer.

reflections between the mirror surfaces. If the light enters at an angle θ and the mirror spacing is d, and assuming that there is no loss at the mirrors, the transmission of the device is given by

$$T = \frac{(1-R)^2}{(1-R)^2 + 4R\sin^2(\delta/2)}, \quad (3.26)$$

where R is the mirror reflectivity and δ, the difference in phase produced by a round trip between the mirror surfaces, is given by

$$\delta = (4\pi d/\lambda)\cos\theta. \quad (3.27)$$

The relationship between T and δ is shown graphically in figure 3.28. When δ is equal to integer multiples of 2π, the phase of all the multiple reflections is the same, and so the transmission will be 100%. The distance between the transmission peaks is called the *free spectral range* (FSR). By using equation (3.27), the FSR can be expressed in units of frequency as $c/(2d\cos\theta)$.

For a collimated beam of light, a Fabry-Perot interferometer has the transmission curve shown in figure 3.28, and hence acts as an optical filter. The ratio of the FSR and the width of the filter $\Delta\nu$, ($\nu_{FSR}/\Delta\nu$), is called the *finesse* of the interferometer and denoted by F. F depends on the mirror reflectivity R and has the functional form $\pi R^{1/2}/(1-R)$. If mirrors with very high reflectivity R are used, then F can become very large, resulting in extremely narrow transmission peaks.

Fabry–Perot interferometers cannot be used in situations when the spectral width of the instrument light is larger than the FSR of the instrument, because different diffraction orders will overlap spectrally. Hence, Fabry–Perot interferometers are only effective when the incident light has a relatively narrow spectral width. However, the resolution of Fabry–Perot interferometers is extremely good. For example, if $d = 10$ mm and $\theta \sim 0°$, the FSR is ν_{FSR}

~ 15 GHz (0.5 cm^{-1}). If the finesse is 100, the spectral resolution is 150 MHz. Hence, it can be seen that Fabry–Perot interferometers have *extremely* high resolution but only over a narrow spectral range.

If the light source has a high degree of divergence, the interference fringes from a Fabry–Perot interferometer will be ring-shaped. The frequency spacing between the rings is v_{FSR}, and so the spectrum of the light source can be determined by measuring the radii of the fringes. A scanning Fabry–Perot can be made by moving one of the mirrors with a PZT element, and thus changing d during the measurement. Another way of achieving the same result is to vary the pressure of the gas inside the device, which also changes the optical path length. Another use of etalons is as elements inside laser cavities, where they act to reduce the bandwidth of the laser spectrum.

3.3.3 Optical Waveguides

There are two main basic ways of transmitting light over distances with low loss: transmission through space, and transmission through an optical waveguide. There has been rapid progress in the field of guided light transmission in recent years, driven by the optical communications industry, and these light guides also are useful in measurement fields. The two types of light guides are slab guides and optical fibres. Slab guides rely on the difference in refractive index between the surface of a dielectric material and a thin layer near the surface. These are used for semiconductor lasers and optical integrated circuits. Optical fibres, the main subject of this section, are the type of guide used in measurement systems.

A detailed description of light propagating through optical fibres can be found in textbooks on electromagnetic wave theory [13,14]. This section will concentrate on the properties of fibres used in measurement applications. The two main types of fibres, shown in figure 3.29 are *step-index fibres* and *graded-index fibres*. In a step-index fibre, the refractive index of the core is slightly larger than that of the outer cladding region. The light propagates along the fibre by reflecting from the boundary between the two regions, as shown in the figure. The angle, θ, for which total internal reflection occurs is given by $\theta = \sin^{-1}(n_2/n_1)$ where n_1 and n_2 are the refractive indices of the core and cladding respectively. Using this equation, and expressing the refractive index in terms of the relative difference in refractive index, $\Delta = (n_1 - n_2)/n_1$, the maximum incident angle, δ_{max}, for propagation can be shown to be

$$\sin\delta_{max} = \sqrt{n_1^2 - n_2^2} \approx n_1\sqrt{2\Delta} \qquad (3.28)$$

$\sin(\delta_{max})$ is called the *numerical aperture* (NA) of the fibre.

Light that is transmitted along an optical fibre travels in modes whose form depends on the incidence angle of the light entering the fibre. If both the core radius and refractive index difference are small, the number of possible modes is greatly reduced. In the limiting case, only the fundamental mode can propagate. This type of fibre is called a single-mode fibre. Single-mode fibres for optical

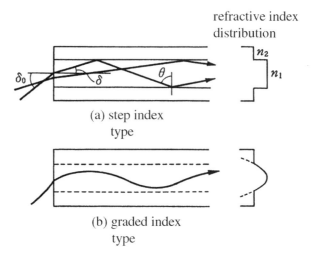

Figure 3.29. Types of optical fibres.

communications have core diameters of 3–10 μm and Δ of the order of 0.2%. Because there is no mode dispersion in a single-mode fibre, this type of fibre is widely used in optical communication. However, great accuracy is required for coupling of single-mode fibres, and for this reason, fibres used for measurement and power transmission have much larger diameters and larger numerical apertures.

Generally, the polarization of light is randomized when it passes through a fibre. However, a special type of fibre has been developed which preserves the light's original plane of polarization. This fibre is called a *polarization conserving fibre*. This fibre has a core whose composition is not symmetric.

The most common kind of optical fibre is made from fused quartz, with a core that is doped with atoms such as Ge and P. In recent years, the transmission loss of quartz fibres have been greatly reduced by purifying the core, and fibres with losses as small as -0.2 dB km^{-1} at $\lambda = 1.5$ μm are available. At longer wavelengths, the absorption of quartz rises, and so the loss increases. At shorter wavelengths, loss due to Rayleigh scattering, which has a λ^{-4} dependence, is the main loss mechanism. At ultraviolet wavelengths, significant amounts of multi-photon absorption and nonlinear optical effects also occur. Laser light from a 308 nm XeCl excimer laser can be transmitted over short distances with few problems, but even this cannot be achieved with light from a 249 nm KrF excimer laser.

In addition to quartz, acrylic materials such as PMMA are used in so-called plastic fibres, and in multi-component glass fibres. These fibres have high loss compared with quartz fibres, and so are not practical for long distance transmission. However, such fibres, with large core diameters of 0.2–1 mm and numerical apertures of up to 0.5, can be used effectively for measurement purposes.

Quartz fibres can be used to transmit light with wavelengths up to about 1.8 μm. In the wavelength region up to 4 μm, fluoride glass fibres can be used, while chalcogenide glass containing species such as AsS and As–Ge–Se can be used for wavelengths up to 6 μm. Also, a multi-crystal halogen fibre is being developed for use with 10 μm CO_2 laser light. A multi-crystal fibre called KRS-5 containing TlBr/TlI has been reported as having a transmission range of 0.5 μm to 40 μm.

3.3.4 Other Optical Elements

A list of optical elements necessary for laser-aided diagnostic techniques is shown in table 3.7. They were classified by the following principle, although the practical distinction between passive elements and active elements is not really very clear. Passive elements are defined as those that change the properties of the light wave, such as propagation direction, intensity, phase and polarization, passively. Active elements are defined as those that use some form of interaction between the light and the element material to change the properties of the light. Such properties can include the wavelength, energy and polarization.

Active elements. Various different active elements are listed in table 3.7. Some of these, such as light sources, nonlinear elements and detectors, are described in other sections of this chapter. This section is a discussion of the remaining categories: modulators, polarizing elements and isolators. Electro-optic, magneto-optic and acousto-optic effects are the basis of many of these elements [14,15].

An *electro-optic* (EO) *effect* can be defined as a phenomenon in which the refractive index of a material is altered by the application of an electric field. The effect is classified as a first-order EO effect, also called a *Pockels effect*, when the refractive index change is proportional to the applied field, and as a second-order EO effect, also called a *Kerr effect*, when the change is proportional to the square of the applied field.

A Pockels cell is a light modulator that uses a peizo-electric crystal such as KDP or $LiNbO_3$. Anisotropic crystals generally show a phenomenon called *birefringence*. In a birefringent crystal, the phase velocity of light passing through the crystal depends on the polarization of the light with respect to a particular crystal axis. When light enters such a crystal, there is a difference in the phase speed of the different components of polarization, and, in general, light will be elliptically polarized on exiting the crystal. Figure 3.30 shows the possible final polarizations of light that entered a crystal polarized at 45° to the x and y directions. The final polarization depends on the difference δ in phase speeds between the light components polarized in the x and y directions. In a Pockels cell, electrodes are attached to the crystal and the value of δ is changed by altering the applied voltage. If δ has a value of π, the polarization will be

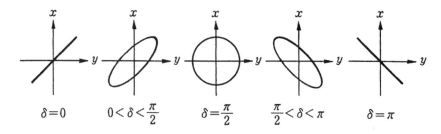

Figure 3.30. Polarization of the wave formed by synthesis of components polarized in the x and y directions. δ is the phase difference between the components.

Table 3.7. Classification of optical components.

Active elements	light sources	laser, light-emitting diode, lamps
	nonlinear optical elements	higher harmonic generation devices, light mixers, optical parametric oscillators
	modulation elements	Pockels cells, ultrasonic wavegenerators, mechanical shutters
	polarizing elements	ultrasonic scanners, EO-polarizers, polygon mirrors, holographic scanners
	detectors	phototubes, photodiodes, photoconductive cells, CCD cameras, streak cameras
	directional elements	Faraday isolators
Passive elements	lenses	spherical lenses, aspherical lenses, cylindrical lenses
	mirrors	metallic-coated mirrors, dielectric-coated mirrors, corner cubes
	optical filters	coloured-glass filters, neutral-density filters, interference filters
	dispersing elements	prisms, diffraction gratings
	polarizers	Glan–Thomson prisms, polarizing filters, polarizing beam-splitters
	waveplates	1/2 waveplates, 1/4 waveplates, Fresnel rhombs, Babinet–Soleil compensators

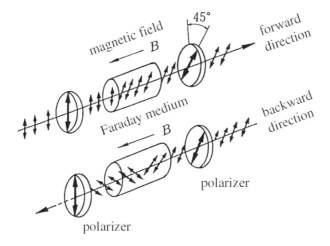

Figure 3.31. Schematic arrangement of an optical isolator.

changed by exactly 90°, and the cell can be combined with a polarizer element to form a fast optical shutter, which can transmit or block the incident light depending on the value of the voltage applied to the cell.

One of the main advantages of Pockels cells is that they have extremely fast response times. Because the response of the crystal usually is very fast, the response time of the cell is determined by the rise time of the voltage pulse applied to the cell. One problem with Pockels cells is that a high voltage, usually at least 1000 V, is required to produce the necessary change in phase speed. Shutters based on Pockels cells have been developed with on–off times of less than a few nanoseconds. Even faster modulation, at frequencies of more than 10 GHz, has been achieved by placing the cell in a microwave waveguide.

The *magneto-optic* (MO) effect includes both the *Faraday effect* and the Cotton–Mouton effect, but only the Faraday effect is of interest for laser diagnostics. This is a phenomenon in which the polarization of light is rotated when it passes through a medium under the influence of a magnetic field. For linearly polarized light, the angle through which the polarization is rotated is given by $RBL\cos\theta$, where R is *Verdet's constant*, B is the magnetic field strength, L is the length of the medium and θ is the angle between the magnetic field and the optical axis. Magneto-optic effects are significant for isotropic materials such as lead glass.

The Faraday effect is used as the basis of optical isolators and magnetic field sensors. An optical isolator is a device that allows light to pass through in only one direction. An example of such a device is shown in figure 3.31. Two polarizers are placed with their polarization axes at 45° to each other, and a Faraday element is placed between them. The change in polarization induced by the Faraday element enables light to pass through the device in one direction, but not in the reverse direction.

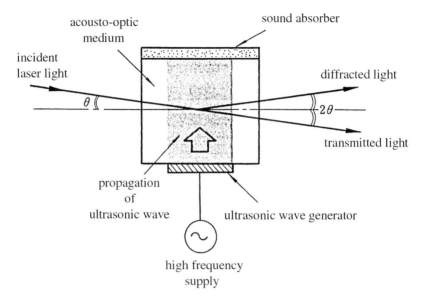

Figure 3.32. Schematic diagram of an acousto-optic cell.

The *acousto-optic* (AO) effect is a phenomenon in which light passing through a medium is refracted by sound waves. This effect is used to control the direction of propagation of light beams. Figure 3.32 shows an example of a device called an acousto-optic cell. Periodic density modulations are generated by an ultrasonic wave in the cell, and light passing through the cell undergoes Bragg scattering due to the density modulations. Thus the propagation direction of the light is altered. Acoustic frequencies in the range of tens of MHz to hundreds of MHz are necessary. When used as a tuning element or a shutter, AO cells have a much slower time response than Pockels cells, but they have the advantages that only a small voltage is required to activate the cell, and repetition rates in the order of kHz are possible. Materials such as $PbMoO_4$ and TeO_2 crystals typically are used for AO cells.

Passive elements. Only a few brief comments will be made here about passive optical elements. For lenses and windows, one of the most important properties is the range of wavelengths over which they can be used. Transmission curves for many optical crystals were shown in figure 3.23, but BK7 glass and quartz glass are the most widely used materials for visible wavelengths. These materials have high transmission from near ultraviolet wavelengths to the infrared. For wavelengths less than 200 nm, MgF_2 and LiF can be used, while no transparent material exists for wavelengths lower than about 100 nm. Materials useful in the infrared region include Al_2O_3, which can be used for wavelengths up to about 6 μm, Ge, up to about 20 μm, KCl, up to about 30 μm, and CsI, which is effective up to about 70 μm.

Mirrors usually are made of a stable material with a metallic or dielectric coating. Metallic coatings deposited by evaporation are inexpensive and widely used, but reflectivity of more than about 95% is difficult to achieve and they are easily damaged by intense laser beams. Mirrors with multilayer dielectric coatings have alternate coatings of dielectrics with different refractive indices, and can have extremely high reflectivity. The same principle can be used to make anti-reflection coatings and interference filters. Compared with mirrors with metallic coatings, mirrors with multilayer dielectric coatings have high reflectivity over a relatively small range of wavelengths, but coatings with reflectivity of arbitrary value, up to close to 100%, can be designed, and coatings with extremely high damage thresholds can be produced. These mirrors are widely used for lasers. The reflectivity depends on the incident angle and the polarization of the incoming light. Dichroic mirrors have high reflectivity only over particular wavelength ranges, and can be used to separate light into different wavelengths.

Anti-reflection (AR) coatings can be made with either a single-layer coating or a multilayer coating. Single-layer coatings can reduce reflection down to about 1.5% while multilayer coatings can reduce it to less than 0.5%. If a coating at only one particular wavelength is desired, reflectivity of 0.2% can be obtained.

Optical elements that use the complex refractive index of crystals are *polarizing prisms* and *waveplates*. Different types of polarizing prisms include the Nicol, Glan–Thomson and Wollaston types. All these types of prisms use reflection at the surface of a prism material containing calcite to separate the light into two polarization components. Separation ratios of 10^{-5} order can be achieved. An alternative polarization element are filters that use polymer coatings but, although inexpensive, these have relatively poor separation ratios.

Waveplates are optical elements that use certain birefringent crystals such as quartz and mica to alter the polarization of light. The change in polarization occurs due to the process shown in figure 3.30. Elements for which the phase shift, δ, is $\pi/2$ are called *quarter-waveplates* while those for which δ is π are called *half-waveplates*. Linearly polarized light that passes through a quarter-waveplate becomes circularly polarized while the effect of a half-waveplate is to rotate the polarization by 90°. Naturally, these waveplates are only effective for particular wavelengths, but elements called *Babinet–Soleil compensators* can produce any given phase change at any given wavelength. Prisms that perform the same function also exist, and these are called *Fresnel rhombs*.

3.4 Detectors and Signal Processing

3.4.1 Light Detectors

The detector is one of the most important components in all types of optical measurement systems. The choice of detector for a particular experiment depends on many factors, such as its absolute sensitivity, noise level, spectral sensitivity and temporal response. Although there are many different types of detectors, they can be grouped into several broad categories depending on the detection principle. Table 3.8 shows optical detectors classified in this way [12,15], and these detectors are described separately below.

Detectors based on photo-electric emission. These types of detectors called *phototubes* consist of a light-sensitive surface called a *photocathode* from which electrons are ejected in response to incident light, and an *anode* at which the electrons are collected. In general, the spectral response of the detector is determined by the type of material used for the light-sensitive surface. The spectral sensitivity of some different materials is shown in figure 3.33. The relationship between the two properties shown in the figure, the sensitivity S and the quantum efficiency Q, is given by $Q = Sh\nu/e$, where ν is the light frequency and e is the electronic charge. As can be seen from the figure, the upper limit of measurable wavelength is about 1 μm, and detection of light down to vacuum ultraviolet wavelengths is possible, although special window materials must be used for these short wavelengths.

Photo-electric emission itself is an extremely fast process, and this can be used as the basis for picosecond streak cameras. However, the sensitivity of phototubes is low, and semiconductor-based detectors gradually are replacing them in many applications. In this class of device, only the biplanar phototube, which has a response of about 100 ps, is a useful device.

The *photomultiplier tube* (PMT) is a form of phototube that contains a set of intermediate electrodes, called *dynodes*, to amplify the photo-electron signal that comes from the photocathode, and hence improve the detector sensitivity. The amplification principle, shown in figure 3.34, is that the original photoelectrons strike the dynodes, and generate secondary electrons at each electrode. This efficient amplification mechanism gives photomultiplier tubes the best sensitivity and the best signal-to-noise ratio of all light detectors. For a detector with applied

Table 3.8. Classification of optical detectors.

Detection principle	detector
Photo-electron emission	phototubes, photomultiplier tubes, streak cameras, image intensifiers
Photovoltaic effects	photodiodes, phototransistors, avalanche photodiodes, CCD cameras
Photoconduction	photoconductive cells
Thermal effects	thermoelectric detectors, bolometers, pyroelectric detectors
Photo-chemical effects	various types of photosensitive films

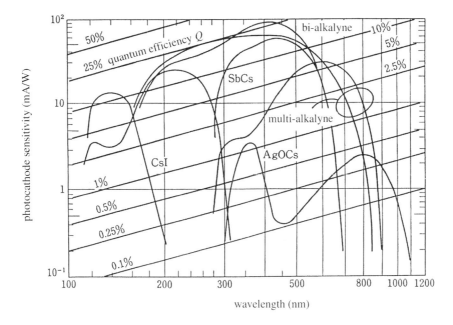

Figure 3.33. Spectral properties of different materials used for photocathode surfaces (data courtesy of Hamamatsu Photonics).

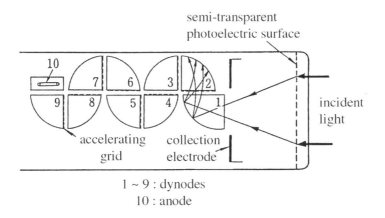

Figure 3.34. Internal configuration of a box-type photomultiplier. The voltage on each dynode gradually increases with respect to the cathode.

voltage E and n amplification stages, the amplification gain is proportional to $\alpha E n$, where α is a constant equal to about 0.7–0.8. The number of dynodes, n, is usually about 10. Gain of up to the order of 10^6 can be achieved.

When the number of incident photons becomes small, the anode current flows in small pulses, where each pulse corresponds to an individual incident

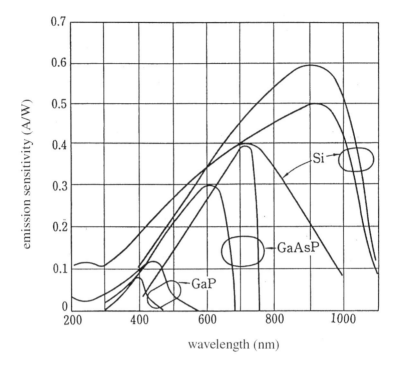

Figure 3.35. Spectral responses of the main types of photodiodes (data courtesy of Hamamatsu Photonics).

photon. In this case, the photomultiplier noise comes from two types of noise sources, *Johnson noise* and *shot noise*. Johnson noise comes from fluctuations in the dark current, and causes spurious pulses in the output which are unrelated to the incident light. Shot noise is due to fluctuations in the electron amplification process. As shot noise increases proportionally with the amplification gain of the detector, it is usually the most important noise source of photomultipliers.

Detectors based on photovoltaic effects. When light falls on a PN junction or on a junction between a metal and a semiconductor, an electron–hole pair is produced, and the movement of these particles in the electric field in the interface region produces an electromotive force. This is the principle of selenium and silicon solar cells. However, the same principle also can be used as a way of detecting light, by applying a reverse bias voltage across the interface and detecting the reverse current that is produced when light hits the surface. This kind of detector is called a *photodiode*.

Figure 3.35 shows the spectral sensitivity of several typical types of photodiodes. Another type, not shown in the figure, is a GaInAs diode that was developed for optical communications at wavelengths up to 2 μm. As can be seen from the figure, Si diodes are most sensitive in the near infrared wavelength region, between 700 nm and 1000 nm. A type of diode having a high resistance I layer in between the P and N layers, called a *PIN diode*, has a very fast temporal response, of down to a hundred picoseconds. Ordinary photodiodes do not have any internal amplification, and so their sensitivity is usually poorer than that of photomultiplier tubes. The most sensitive type of diodes are *avalanche photodiodes*, in which amplification is achieved by applying a strong reverse bias voltage and relying on collisions of the carriers in the depletion layer. Although the amplification gain of these devices is, at most, a few hundred, which is much less than that of a photomultiplier, quantum efficiency can be as high as 70–80%. These have been used as detectors for Thomson scattering experiments that use the fundamental wavelength of a Nd:YAG laser (1.06 μm). In addition, a very fast avalanche photodiode, with response times of less than 100 ps, also has been developed for optical transmission application.

The noise level in semiconductor light detectors generally is stated in terms of a quantity called the *noise-equivalent power* (NEP). This noise figure corresponds to the amount of incident light power that would be necessary to produce a signal-to-noise ratio of 1, and can be written as

$$\text{NEP} = P\left(\frac{V_n}{V_s}\right)(\Delta f)^{-1/2} \quad (\text{W Hz}^{-1/2}). \tag{3.29}$$

In this expression, P, V_s, V_n and Δf are the values of the incident power, signal voltage, noise voltage and equivalent bandwidth, respectively. As the noise usually is proportional to the detection area, the NEP level usually can be improved by reducing this area. Also, for this reason, the noise sometimes is expressed in the form of $D^* = $ (receiving area)$^{1/2}$/NEP. For the case of a Si photodiode, the NEP is about 10^{-14}–10^{-15} W Hz$^{-1/2}$. For semiconductor detectors, noise sources also include thermal noise and shot noise. In addition, in nearly all cases, the diode is used with an electric amplifier, and the amplifier noise also has to be considered.

Detectors based on photoconduction effects. In the photoconduction effect, the resistance of a semiconductor is changed by incident light. One example of a detector based on this effect is the CdS cell, which is widely used as a detecting element in cameras and electrical devices because it is inexpensive, works well at visible wavelengths, and has sensitivity close to that of the human eye. However, it has a relatively slow response time (μs order) and poor linearity, so it rarely is used in spectroscopy applications. The importance of photoconductive detectors arises from the ability to detect light of long wavelengths, more than 1.5–2 μm, for which photomultipliers and photodiodes cannot be used.

Figure 3.36 shows the relative detection sensitivity of photoconductive cell detectors. PdS and PbSe detectors have good response at wavelengths up to about 5 μm if they are cooled with liquid nitrogen. For wavelengths longer than this, materials such as InSb, HgCdTe and Ge:Cu can be used: all these materials also require cooling. Response times of the order of 1 μs usually are achievable, depending on temperature and materials.

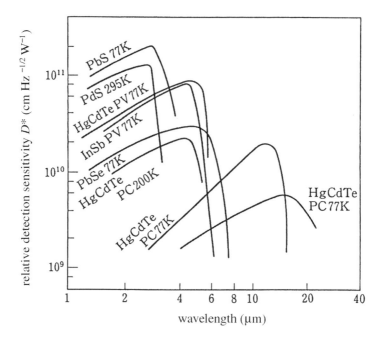

Figure 3.36. Spectral responses of the main types of infrared photoconduction detectors.

Thermal detectors. Thermal detectors almost always have poorer sensitivity and lower response than photoelectric type detectors, but they have the advantages that they have no wavelength dependence and that the absolute sensitivity can be calibrated easily. They are widely used for light detection at infrared wavelengths, where there are few good detectors, although even there they gradually are being replaced with photoconduction detectors. In the field of laser-aided diagnostics, the use of thermal detectors is restricted to measurement of absolute laser power or energy. Different detectors exist for measuring the power of cw and pulsed lasers. For the measurement of pulsed laser energy, the maximum temperature of the detector surface is recorded and then converted to a value of energy in joules.

The detectors can be separated, on the basis of measurement principles, into thermoelectric detectors, which use thermocouples or thermopiles, and pyroelectric detectors, which detect the change in self-polarization induced by temperature changes in the material. The latter type has a faster time response, but cannot be used for measurement of cw light.

3.4.2 Imaging Detectors

The ability to obtain large amounts of spatially resolved information simultaneously is one of the key advantages of optical measurement techniques.

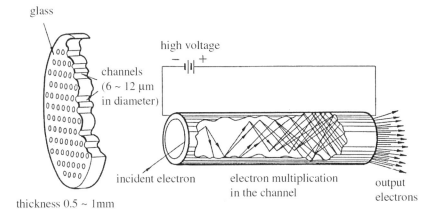

Figure 3.37. Amplification principle for a micro-channel plate.

In the past, cameras were used as detectors for optical measurements when one- and two-dimensional information was desired. Recently, however, there has been a rapid development of photoelectric devices, and these detectors have greater sensitivity than cameras. In addition, the image usually is obtained in digital form, which enables further analysis by image processing techniques.

An *optical multichannel analyser* (OMA) is an instrument that combines a monochromator and a one-dimensional array detector so instantaneous spectra can be measured. A photodiode array with a shift register is used, and the output signals are automatically read by a computer. A typical array has dimensions of 20–50 mm and pitch of $\sim$50 μm.

An *image intensifier* (II) is a device that uses the electron multiplication effect of a microchannel plate to increase signal intensity. This can be placed in front of an array detector to improve the sensitivity. Figure 3.37 shows the amplification principle of a microchannel plate. These devices have an amplification factor of up to 4000, and can be gated with extremely short voltage pulses. Image intensifiers also can be attached to cameras, and provide a significant increase in detection sensitivity.

CCD cameras have become the most widely used instrument for two-dimensional imaging. In these devices, charge is accumulated in potential wells formed inside semiconductors. By changing the externally applied voltage, the charge can be moved from cell to cell in turn, enabling the signal from each pixel to be read out. Devices with pixel arrays as large as 2048 × 2048 have been developed. By cooling the individual cells, in order to reduce thermal fluctuations in the size of the accumulated charge, sensitivity down to photon counting levels can be achieved. Recently, digital cameras based on a CCD array of more than 200 million pixels have become available cheaply. Infrared imaging cameras with combinations of infrared detectors and scanning units also have been developed.

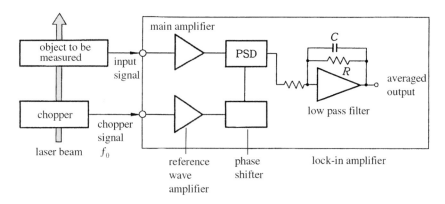

Figure 3.38. Schematic diagram of a lock-in amplifier.

3.4.3 Noise Sources and Signal Recovery

The electrical signal obtained from optical detectors always contains an unwanted noise component. It usually is possible to remove the noise, or at least reduce its effect, if there is an intrinsic difference between the nature of the noise and that of the original signal. There are several different kinds of noise. One type, called white noise, has no frequency dependence, is distributed over a very wide frequency range, and usually can be time-averaged to zero. Another type of noise is cyclic noise, which may or may not be synchronous with the measured signal. Synchronous cyclic noise is difficult to remove because it usually does not average to zero over time, and it cannot be differentiated from the real signal with Fourier analysis.

In this section, we will describe instruments and methods that are used to recover signal from both white noise and synchronous cyclic noise.

Lock-in amplifiers. A schematic diagram of a lock-in amplifier is shown in figure 3.38. The measured signal is modulated at a low frequency of f_o by a device such as a chopper, and a *phase-sensitive detector* (PSD) inside the lock-in amplifier extracts only the part of the measured signal that is synchronized with f_o. This eliminates white noise, noise at other frequencies, and noise not synchronized with the signal. A phase-sensitive detector is a type of multiplexer between the signal and reference waves. After this step, the signal is multiplied with a square reference wave in order to remove its DC component. When the reference wave is square, its Fourier transform includes odd harmonics, and although a DC component is not generated by the multiplication of these odd harmonics and signal, a DC component is generated by odd harmonics in the signal itself. The synchronous amplifier shown in figure 3.38, which is a narrow-band amplifier, removes these harmonics. The low-pass filter in the output end of the circuit ensures that the output signal only contains the component of the input signal that was synchronized with f_o.

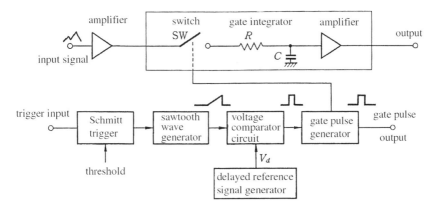

Figure 3.39. Schematic diagram of a boxcar integrator.

Boxcar integrators. A schematic diagram of a boxcar integrator is shown in figure 3.39. This device can be used to detect a signal such as that produced in a pulsed laser measurement, where the signal is large only for a short time. The circuit that produces the sampling pulse is triggered synchronously with the signal input, and only the portion of the signal that exists during the sampling window (i.e. during the time of the sampling pulse) is transmitted through the circuit. This output is then time-averaged. In a laser-aided diagnostic experiment, a simple way of generating the trigger pulse required for this device is to use a photodiode to detect a part of the laser beam.

A boxcar integrator is effective at reducing the effect of noise and thereby improving the signal-to-noise ratio because only the part of the signal with a high signal-to-noise ratio is averaged. Another way to use this device is to slowly scan the sampling window through the entire signal pulse. In this case, the output waveform is similar to that produced by a fast sampling oscilloscope. The time resolution of boxcar integrators is determined by the width of the sampling window, and commercial devices with window times as short as 2 ps are available.

Photon counters. A schematic diagram of the arrangement used for a photon counter based on a photomultiplier tube detector is shown in figure 3.40. This type of device can be used to improve the signal-to-noise ratio in experiments in which the signal consists of a small number of photons arriving randomly at the detector. The detected electrical signal consists of pulses due to individual photons and pulses due to dark noise. The intensity of the dark noise pulses, however, usually is smaller than the real signal pulses, and so a discriminator can be used to effectively reject these noise pulses.

A further improvement can be obtained by modulating the input light signal with a chopper and using an up/down counter to add any pulses that

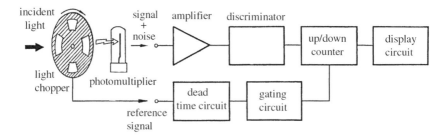

Figure 3.40. Schematic diagram of a photon counting instrument.

arrive when the chopper is open, and subtract any pulses that arrive when it is closed. In this way, the noise pulses that arrive when the chopper is opened are compensated for.

3.4.4 Observation of Fast Waveforms

There are many measurements in which it is necessary to measure pulsed waveforms with very short pulse times. The temporal resolution of detection systems in optical experiments is determined by the temporal responses of both the light detector and the signal processing system.

The limit to the temporal response of photomultiplier tubes is determined by fluctuations in the time it takes for photo-electrons to move through the dynode system. This time is of the order of 2–10 ns. In a microchannel plate, the distance is much shorter, and response times of 300 ps order are possible. In a biplanar phototube, the photocathode and the anode are parallel and close together, and this detector has a response time of about 100 ps. Photodiodes with very large reverse bias voltages can be used to achieve response times of 50 ps.

The fastest optical detector is the *streak camera*, for which response times of 0.4 ps and sensitivity at photon counting levels is achievable. Such a device is shown schematically in figure 3.41. The light that passes through the entrance slit strikes the photocathode, producing photo-electrons. These electrons are accelerated by a high voltage grid, and then pass between two parallel electrodes, are amplified by a microchannel plate and then detected by a device such as a CCD camera. Fast time response can be achieved by applying a ramp voltage across the electrodes rapidly to scan the electron beam. The key to the fast time response of the system is the parallel electrodes. The voltage across the electrodes is rapidly scanned during the measurement so that the spatial position at which an electron strikes the microchannel plate depends on the time that the electron passed through the electrodes. In other words, the spatial position of the signal on the final image indicates the time that the light entered the detector.

This arrangement can be used to measure single pulses, and averaging over repeated signals improves the signal-to-noise ratio. It also is feasible to observe the distribution of light across the entrance slit, thereby obtaining the one-

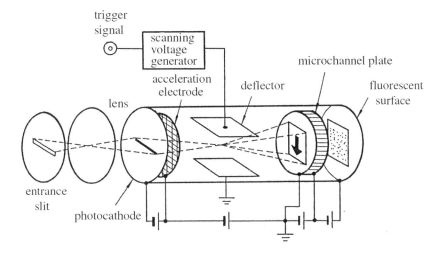

Figure 3.41. Schematic diagram of a picosecond streak camera.

dimensional spatial dependence in addition to the temporal dependence of the signal.

In general, oscilloscopes are used for directly recording waveforms. In recent years, digital oscilloscopes with memories for recording waveforms have become used commonly in most laboratories. For the cases when the signal is not in the form of a single pulse, but is repeated at some repetition frequency, sampling oscilloscopes can be used. For single pulse detection, instruments with bandwidths up to 1 GHz have been developed, while sampling oscilloscopes can be used for up to about 40 GHz. The relationship between the bandwidth F (MHz) and the effective rise time T_r (ns) is $T_r = 350/F$. Furthermore, in a digital memory oscilloscope, the recording speed W_s, which can be considered to be the response time that includes the effect of the wave amplitude A, is given by

$$W_s = 800A/T_r = 2.28AF \text{ (cm } \mu\text{s}^{-1}) \tag{3.30}$$

where A is in units of centimetres. For example, for a waveform with $A = 1$ cm and $T_r = 1$ ns, an oscilloscope with bandwidth of 350 MHz and recording speed of 800 cm s^{-1} is necessary.

Measurement of extremely fast optical phenomena, on femtosecond-order timescales, is inherently difficult using the type of electronic devices described here. One possible method, not discussed here, is to use the nonlinear correlation of second-harmonic generation and two-photon absorption processes.

REFERENCES

[3.1] Young M 1977 *Optics and Lasers* (Springer-Verlag)

[3.2] Yariv Y 1967 *Quantum Electronics* (John Wiley & Sons)
[3.3] Shimoda K 1984 *Introduction to Laser Physics* (Springer-Verlag)
[3.4] Maeda M 1987 *Quantum Electronics* (Shokodo) (in Japanese)
[3.5] Shapiro S L ed 1977 *Ultrashort Light Pulses* (Splinger-Verlag)
[3.6] Demtröder W 1981 *Laser Spectroscopy* (Splinger-Verlag)
[3.7] Maeda M 1984 *Laser Dyes* (Academic)
[3.8] Moulton P F 1985 *Laser Handbook* vol 5 ed M Bass and M L Stitch (North-Holland) p 203
[3.9] Lin T 1990 *Opt. Quantum Electron* **22** 283
[3.10] Simon U and Tittel K F 1994 *Laser Focus World* p 99
[3.11] Born M and Wolf E 1974 *Principles of Optics* (Pergamon)
[3.12] Yoshinaga H ed 1973 *Handbook of Applied Spectroscopy* (Asakura)
[3.13] Marcuse D 1974 *Theory of Dielectric Optical Waveguides* (Academic)
[3.14] Sueta T 1985 *Opt-Electronics* (Shokodo) (in Japanese)
[3.15] Yariv Y 1971 *Introduction to Optical Electronics* (Holt, Rinehart and Winston)

PART II

APPLICATIONS AND MEASUREMENTS

Chapter 4

Plasma Measurements

Laser-aided plasma diagnostics are one part of a large group of diagnostic methods for plasmas that involve the measurement of electromagnetic waves. This large group of measurement methods, collectively known as 'plasma spectroscopy', has been studied and developed for many years. Laser-aided plasma diagnostics, together with microwave-based methods, form a subset of this large group which can be called 'active plasma spectroscopy'.

Plasma diagnostic measurements involving the measurement of electromagnetic waves can be classified in four ways according to the range of wavelengths that is used, the experimental technique that is used, the plasma information that is obtained, and the plasma phenomena that are used to gain the plasma information. With these four categories in mind, the material in this chapter is presented as follows. Firstly, section 4.1 is an overview of the measurement of emission from the plasma, known as passive plasma spectroscopy. This technique is strongly related to all laser-aided methods. The next two sections contain descriptions of laser-aided measurements, firstly in high-temperature plasmas, in section 4.2, and then in various types of low-temperature discharge plasmas, in section 4.3.

4.1 Overview of Plasma Spectroscopic Methods

There are a large number of methods that determine plasma properties by measurement of electromagnetic waves emitted by various species in the plasma. Measurement techniques based on wavelengths in the visible or near visible part of the spectrum have been developed since the beginning of this century, and are powerful techniques for the study of plasmas [1]. In relatively low temperature plasmas, with temperatures of less than a few tens of thousands of degrees, many methods have been applied to different plasmas, and the techniques are well understood and well established. High-temperature plasmas that have temperatures of a few tens of thousands to a hundred million degrees have been developed for the goal of controlled thermonuclear fusion. In these high-temperature plasmas, many-electron atoms are multiple-ionized, and emission from these species, together with bremsstrahlung from electrons, forms the bulk of the emitted radiation, at wavelengths ranging from the vacuum ultraviolet to the x-ray part of the spectrum. In the latter wavelength range, measurement instruments and calibration methods still are being actively developed. In addition, in magnetically confined plasmas, there is cyclotron emission in the near millimetre wavelength range. This emission is a useful source of information for determination of the electron temperature. These subjects are summarized in turn in the subsections below.

Identification of spectral lines. Line emission spectra are widely used for identification of species because each particular species of excited atoms, molecules and ions has its own characteristic line spectrum. The range of wavelengths over which emission occurs is extremely wide. In glow and arc discharges, there is strong emission in the visible and infrared part of the spectrum while multiple-ionized ions in high-temperature plasmas emit in the ultraviolet and x-ray wavelength regions.

Spectral line intensity. In a plasma with temperature T, a spectral line corresponding to a transition from energy level l to energy level m has intensity, I_{lm}, that is given by [2]

$$I_{lm} = \frac{A_{lm} g_l}{U} h\nu_{lm} n \exp\left(-\frac{E_l}{\kappa T}\right). \tag{4.1}$$

In this equation, A_{lm} is the transition probability, g_l is the degeneracy of the upper level, U is the internal partition function for the emitting species, ν_{lm} is the frequency of the emitted radiation, n is the density of the emitting species and E_l is the energy of the upper level.

The relationship between I and T can be used as the basis of a measurement technique for T. Figure 4.1 shows the result of an emission measurement in which the temperature was determined using equation (4.1) [3]. In this figure, the horizontal axis is E_l, while the vertical axis is the logarithm of the emission intensity divided by the constants on the right-hand side of equation (4.1). Each data point is labelled with the wavelength λ_{lm} of the emission, where $\lambda_{lm} = c/\nu_{lm}$ (c is the speed of light). As can be seen from the figure, the measured data lay on a straight line, which means that the density distribution amongst the excited states could be well fitted by a Boltzmann distribution. The gradient of this line corresponds to the temperature T.

A simpler, but less rigorous method of temperature measurement (less rigorous in the sense that a Boltzmann distribution is assumed from the start) is to use the ratio of emission intensities of transitions in the same emitting species. This ratio is shown in the next equation.

$$\frac{I_{lm}}{I_{l'm'}} = \frac{A_{lm} g_l}{A_{l'm'} g_l'} \frac{\nu_{lm}}{\nu_{l'm'}} \exp\left(-\frac{E_l - E_{l'}}{\kappa T}\right). \tag{4.2}$$

In one sense, the technique shown in figure 4.1 can be understood as being an extension of this simple method, in that the relative intensities of many spectral

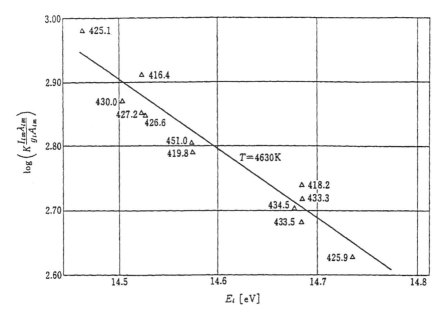

Figure 4.1. Determination of temperature from relative intensities of argon emission lines, in arc plasma at 1 atm. The points show the intensities I_{lm} of Ar I emission lines, and K is a constant. A temperature of $T = 4630$ K was determined from the gradient of the line [2].

lines are used to improve the accuracy of the determination of T.

For plasmas not in thermal equilibrium, the temperature determined from equation (4.1) might be called the *excitation temperature*, because it is the temperature which characterizes the population distribution among excited levels. This temperature usually is related to the *electron temperature* that characterizes the electron energy distribution function (EEDF). If these emission intensities are to be used to properly understand the plasma state, however, a model of the emission process such as a collisional-radiative model or a corona model must be used. A good example of such a technique is the method used to estimate the concentration of impurity species in high-temperature plasmas.

Spectral line shape [1,2]. The emission whose intensity I_{lm} is defined in equation (4.1) does not have an infinitely narrow spectral width, but rather has a finite width determined by the variety of different factors that were discussed in section 2.3. It is possible to obtain information about the plasma from this spectral shape. However, as there are a large number of mechanisms that influence the spectral line shape, it is first necessary to identify which effects are important for the particular plasma being studied.

Several different properties can be determined from spectral line shape measurements, including ion temperature, electron density and plasma current. Ion temperature can be determined from the Doppler-broadened profile of spectral lines in relatively low-density, high-temperature plasmas. In relatively high-density, low-temperature plasmas, microfields due to charged particles in the plasma produce a Stark effect that changes the spectral line shape and the electron density can be determined from this effect. In plasmas with large applied magnetic fields, Zeeman effects affect the spectral line shape, and when plasma current modifies the applied magnetic field, this current can be inferred from the modified spectral line shape.

Bremsstrahlung. The measurement techniques mentioned above all are based on measurement of light that is generated when electrons move between energy levels inside atoms, ions or molecules. Emission also is produced when free electrons are decelerated due to interactions with ions in the plasma. This radiation is spectrally continuous and is called bremsstrahlung. Because bremsstrahlung is caused by the kinetic motion of the electrons, the electron temperature can be determined by the spectral shape of the emission, and the electron density and effective charge state of the decelerating ions can be inferred from the radiation intensity. In some instances, the decelerating electrons might be captured by ions. This produces a continuous spectrum of emission with a band end. As with bremsstrahlung, various plasma properties can be determined from this emission.

Cyclotron emission. In high-temperature plasmas that have large applied magnetic fields, charged particles move in circular motion around the magnetic field lines. This is called cyclotron motion, and the radiation that is emitted due to the accelerated motion of the charged particles is called cyclotron emission. Electron cyclotron emission is much larger than ion cyclotron emission because of the much smaller mass of the electron. In the magnetically confined plasmas used for fusion research, the magnetic field is usually of the order of a few tesla, and the cyclotron frequency, given by equation (1.6), is of the order of 100 GHz. This corresponds to wavelengths in the millimetre wave region. At these relatively long wavelengths, the plasma can become optically thick and radiate like a black body with its temperature characterized by the electron temperature. In these circumstances, the electron temperature can be determined by measuring the emission spectrum. In plasmas with spatially varying magnetic fields, measurement of the cyclotron emission spectrum can provide spatially resolved information about the electron temperature because the intensity of each particular frequency corresponds to the electron temperature at the position that frequency is the electron cyclotron frequency. This principle was developed into an effective measurement instrument in 1975 [4].

Total radiation intensity. In plasmas with temperatures of the order of several million degrees, a few tens of per cent of the total power supplied for plasma generation will be emitted as electromagnetic waves. Most of this radiation has wavelengths in the vacuum ultraviolet and x-ray parts of the spectrum, in the form of line spectra and continuum radiation from bremsstrahlung. An instrument called a *bolometer*, which measures the radiation intensity integrated over all wavelengths, can be used to measure the total radiation intensity. In high-temperature plasmas, this quantity is used for power balance calculations.

4.2 Laser-Aided Measurements in High-Temperature Plasmas

The aim of high-temperature plasma research is to achieve temperatures of about 100 million degrees (10 keV) in order to generate energy by nuclear fusion. The two main schemes, magnetic confinement and inertial confinement, cover a large range in electron density. In inertial confinement plasmas, the electron density is about 10^{30} m^{-3}, and the cut-off frequency for radiation is 10^{16} Hz, which corresponds to wavelengths in the range of 30 nm, in the soft x-ray part of the spectrum. Hence, laser diagnostics cannot be used to measure properties in inertial confinement fusion research, except for the outer region of the core called the *halo*, in which the electron density falls rapidly to zero in a distance of a few millimetres. However, laser diagnostics are very well suited for measurements in the plasmas used in magnetic confinement research. These measurements are discussed in this section.

In order to understand plasma behaviour in magnetically confined high-temperature plasmas, several different types of measurement are required. Firstly, information about the temperature and density of electrons and ions in hydrogen plasmas is required, including plasmas operated in deuterium and tritium. Secondly, information about the density and temperature of neutral particles and any impurity species is needed. Information about velocity distribution functions sometimes is required, because these might differ significantly from Maxwellian distributions in some cases. Finally, information about electric and magnetic fields in the plasma, as well as fluctuations in these quantities, is desirable. These three areas are discussed separately below.

4.2.1 Measurement of Plasma Density and Temperature

The most reliable methods of measuring plasma density and temperature are laser interferometric and scattering methods. Each method has well known advantages and disadvantages, and the methods are used complementarily.

Interferometry. The basic principle of interferometry was outlined in section 2.2. The most common types of interferometers are the Mach–Zehnder type, shown in figure 2.10, and the Michelson type, shown in figure 2.11.

The best and most sophisticated example of an interferometer used for measurements in high-density plasmas is the instrument used on the TEXTOR tokamak at Jülich, Germany [5]. This system is shown in figure 4.2. The laser source is a HCN laser operating at $\lambda = 337$ μm, and fringe shifts from nine channels are measured simultaneously. Because interferometry measures the phase difference due to the total path through the plasma, spatially resolved local information can be obtained only by measuring several chords through the plasma, assuming cylindrical symmetry or some form of density profile along

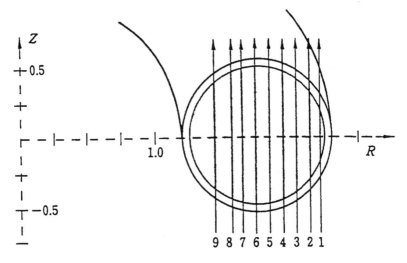

Figure 4.2. Beam positions for the interferometry/polarimetry experimental system on the TEXTOR tokamak. The scales are in metres.

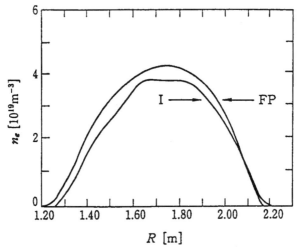

Figure 4.3. An example of an electron density profile in the TEXTOR tokamak measured using interferometry. The I and FP profiles are the results obtained using two different Abel inversion techniques.

the chords, and then using an Abel inversion procedure. This procedure was described earlier in section 2.2. Figure 4.3 shows an example of a spatial profile of electron density in the TEXTOR tokamak obtained using this method. This same instrument also is used to measure the rotation of the laser polarization due to the Faraday effect of the plasma magnetic field. Measurements made using this effect, also discussed in section 2.2, will be described in section 4.2.3.

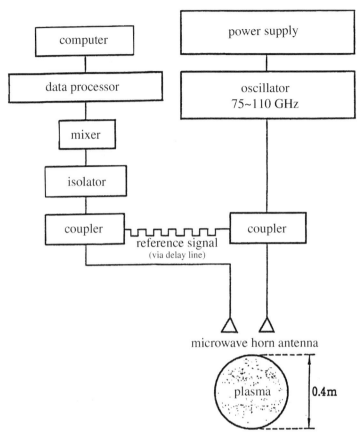

Figure 4.4. Diagram of the microwave reflectometer arrangement for the TFR tokamak. The microwave frequency is swept in the range 100–75 GHz over a 5 µs period, and phase signals from the reflectometer during this time are detected. An Abel inversion method in phase space is used to determine the electron density profile from these results.

As can be seen from figure 4.2, accurate measurement of density profiles using interferometers requires many laser channels across the plasma cross-section. In future fusion devices, however, it will not be possible to have the same kind of optical access, with large numbers of windows, because of the presence of blanket and shield layers that are required to absorb neutrons. To overcome the problem of limited optical access, the instrument shown in figure 4.4, called a reflecto-interferometer, or more simply a *reflectometer*, has been developed.

As discussed in section 2.2.1, the principle of this instrument is that electromagnetic waves are reflected in the plasma at the cut-off layer, and the position of the cut-off layer depends on the frequency of the electromagnetic wave. An interference signal is generated between the beam that is reflected

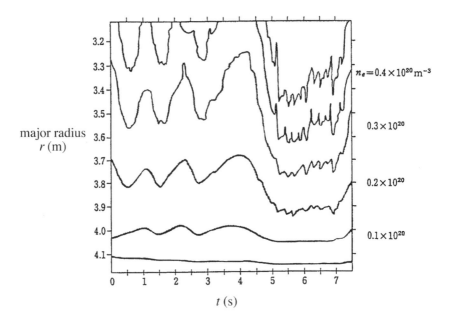

Figure 4.5. Measurement results from the reflectometer system on the JET tokamak. The figure shows the electron density as functions of radius and time. The change in profile during the time 5–7 s, when additional plasma heating was used, can be clearly seen. In addition, variations on a 0.5 s timescale due to sawtooth fluctuations can be seen.

from the plasma and a reference signal, and the phase difference between these signals depends on the distance the beam has propagated into the plasma and the density profile along the propagation path. From these phase data at different frequencies, the spatial distribution of density can be obtained by using an Abel inversion procedure in the frequency space. This technique was demonstrated in principle on the TFR tokamak in France in 1984 [6], and then developed further into a practical measurement technique at the JET tokamak [7]. Figure 4.5 is an example of data obtained by the JET reflectometer [8], showing density contours as functions of time and position.

An instrument that is capable of measuring both density and its fluctuations is currently being developed. Density fluctuations are considered to be one of the main causes of anomalous transport problems in magnetically confined plasmas, and so this new instrument, called a *correlation reflectometer*, promises to be useful for gaining further understanding of high-temperature plasma properties.

Thomson scattering. The principle of laser Thomson scattering was described in section 2.2.4. There are two regimes of scattering experiments, depending on the relative sizes of the laser wavevector k and the Debye length λ_D. When $|k|\lambda_D \gg 1$, the regime is called incoherent Thomson scattering. The regime for $|k|\lambda_D$

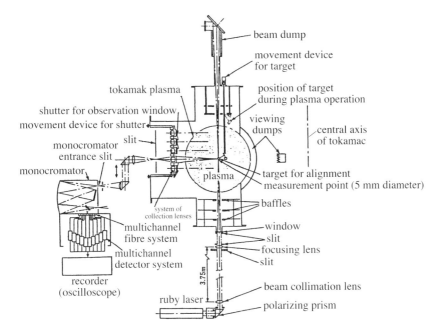

Figure 4.6. Arrangement for the Thomson scattering experiment on the T-3 tokamak. The baffles, viewing dump and beam dumps shown in the figure were for reducing the amount of stray laser light at the 694.3 nm ruby laser wavelength.

< 1 is called collective Thomson scattering. These are described separately below.

Incoherent laser Thomson scattering is the most reliable method for measurement of electron temperature and density in plasmas. The basic principle of this technique is that the electron density is determined from the scattered light intensity and the temperature is determined from the scattered light spectrum.

The most commonly used commercially available laser for this type of incoherent scattering technique is the ruby laser, because of its high power, and efficient spectroscopic and detection devices that are available for it. The most well known example of an incoherent Thomson scattering experiment is the measurement performed on the USSR's T-3 tokamak in 1968 [9]. The arrangement for this experiment is shown in figure 4.6. In addition to the light scattered from the plasma, a significant amount of the incident laser light was scattered from windows, lenses and the chamber walls. Combinations of slit apertures, and beam and viewing dumps were used to minimize the amount of this stray laser light that entered the detection system. In order to measure the spectrum in a single laser shot, the output of the spectrometer used to analyse the scattered light was measured by a multichannel detection system. An example of measured data is shown in figure 4.7. From theoretical curves fitted

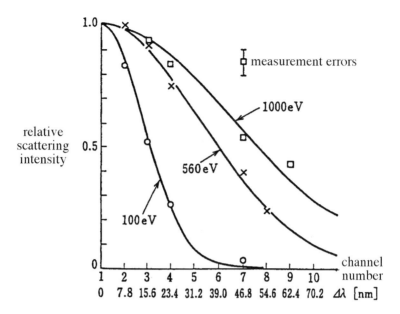

Figure 4.7. Results of the Thomson scattering experiment on the T-3 tokamak. Measurements for plasma currents $I = 85$ kA($\square$), 60 kA($\times$) and 40 kA($\bigcirc$) are shown.

to the experimental data, performed using the procedure explained in section 2.2.4, the electron temperatures for each case were determined to be 100, 560 and 1000 eV.

In the 1960s, lack of progress in fusion research due to anomalously large energy loss across magnetic field lines had led to pessimism over the prospects for controlled thermonuclear fusion. This Thomson scattering measurement, however, demonstrated convincingly that the tokamak device, developed in the USSR, could generate high-temperature, high-density plasmas. After this result, fusion research all over the world shifted to tokamak-based research. In that sense, this measurement was a turning point in fusion research. In the decades since this experiment, further research in many countries has produced tokamak devices in which the break-even condition, the condition at which the amount of energy used to generate the plasma is the same as the fusion energy generated by the plasma, has been satisfied. Another significant consequence of the T-3 Thomson scattering measurement was that plasma researchers in general recognized the usefulness and the importance of laser diagnostic methods. In this sense, the Thomson scattering experiment on T-3 was epoch-making, and its importance can hardly be overstated.

Incoherent Thomson scattering experiments have progressed significantly since the T-3 measurement. The most advanced measurement apparatus yet developed is the TVTS (television Thomson scattering) system developed at Princeton University in the USA [10]. This system, installed on the TFTR

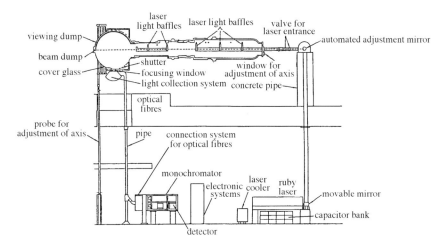

Figure 4.8. Arrangement of the TVTS system on the TFTR tokamak. The chamber cross-section is shown in the upper left part of the figure. The laser beam passes across the plasma cross-section, and scattered light from 76 points in the beam is sent via optical fibres to the detection system.

tokamak, is shown in figure 4.8. In the TVTS system, the scattered signal from many points along the laser beam is collected with a large collection lens and focused into a multichannel optical fibre system. The fibres lead to a spectrometer, produce a two-dimensional output that is detected using an image-intensified CCD camera. The system is arranged so that the signal in one direction corresponds to spatial position in the plasma while the other direction corresponds to wavelength. The signal from 74 different positions along the laser beam can be detected, with spatial resolution of about 10 mm. Figure 4.9 is an example of the data obtained with this system, and shows how the good spatial resolution allows fine plasma structures due to magnetic island formation to be seen.

The simple principle of incoherent Thomson scattering means that very reliable data can be obtained. However, this measurement has the drawback that, in addition to the ports needed for laser injection and signal observation, separate large ports are needed for the beam dump and the viewing dump that reduce the amount of stray laser light. In order to overcome these access problems, a new type of Thomson scattering called LIDAR (light detection and ranging) Thomson scattering was conceived at Stuttgart University in Germany and then applied to the JET tokamak [11]. This system is shown in figure 4.10. The laser source is a ruby laser with pulse width of 300 ps, output energy of 2 J and repetition rate of 0.5 Hz. The scattering angle is 180° and the same port is used for injection of the laser beam and observation of the scattered signal. This arrangement sometimes is called backward scattering. The spatial resolution is determined by the laser pulse width and the response of the detection system, in

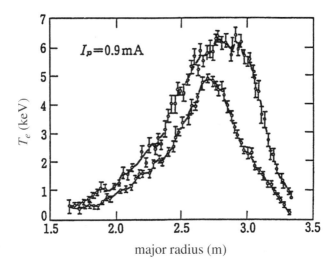

Figure 4.9. Two examples of an electron temperature profile in the TFTR tokamak measured using the TVTS system.

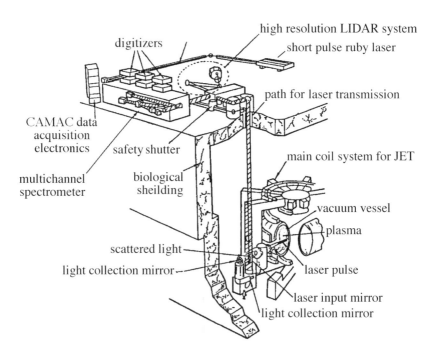

Figure 4.10. Arrangement of the LIDAR measurement system on the JET tokamak.

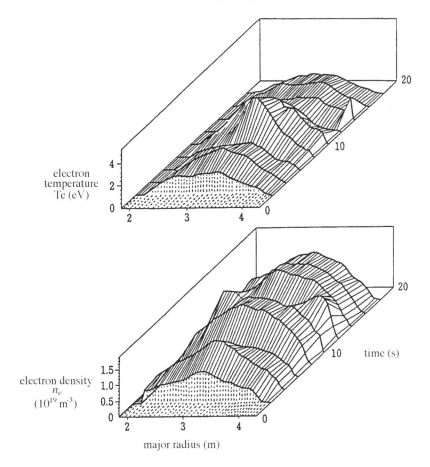

Figure 4.11. Examples of electron density and temperature profiles measured using the LIDAR measurement system on the JET tokamak. The temporal development of these profiles is shown.

the same way as for radar. In this system, the detector is a microchannel plate photomultiplier with response time of 700 ps, and the spatial resolution is about 100 mm. The problem of stray light from the entrance window and other sources is eliminated because stray light arrives at the detector at a different time from the scattered signal.

The LIDAR system on the JET tokamak started working in 1987, and figure 4.11 shows an example of the type of data that now are measured routinely [11]. Since the original system was designed, the photomultiplier detector has occasionally been replaced with a streak camera with temporal response of 300 ps. This improved the spatial resolution to 50–70 mm [12]. In addition, the repetition rate of the laser was improved to 4 Hz, and so data now is taken at 0.25 s intervals.

Another problem of the Thomson scattering method is that very high-

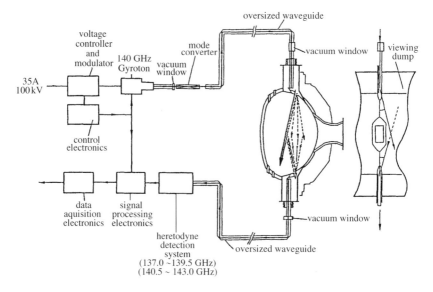

Figure 4.12. An example of an arrangement for a gyrotron scattering experiment on the JET tokamak.

power laser pulses, of the order of 0.1–1 GW, are necessary for a sufficiently large signal to be detected. Such lasers exist, but their repetition rate is not high and so information about temporal changes in the plasma density and temperature cannot be obtained. Continuous monitoring of electron temperature and density is possible using electron cyclotron emission, discussed in section 4.1, and interferometry, discussed in section 4.2.1. Therefore, Thomson scattering now is viewed as a means to provide a very reliable method of calibration for these methods, both in absolute value and in spatial profile, although the comparison can be made only at discrete times.

The measurement systems discussed above can be used to directly measure electron properties with a high degree of accuracy and reliability. However, it also is possible in principle to measure ion properties such as ion density and temperature in a similarly reliable way. As discussed in section 2.2.4, Thomson scattering measurements in the collective regime ($k\lambda_D < 1$) can provide information about T_i in addition to T_e. Several years ago, a study was performed with the aim of evaluating different experimental systems for this measurement in a plasma with $T_e \sim 10$ keV and $n_e \sim 10^{20}$ m^{-3}. Two scattering systems were identified as promising candidates. One of these is based on a cw high-power gyrotron, and requires a scattering angle of 20–30°. This was tested on the JET tokamak in the UK using the arrangement shown in figure 4.12 and first results have been obtained [13]. The second is based on a CO_2 laser ($\lambda = 10.6$ μm) and requires a scattering angle of less than 1°. Such a system is being prepared for application to the JT-60U device at the Japan Atomic Energy Research Institute.

In a deuterium–tritium fusion plasma, alpha particles, which are nuclei of He^4, will be produced in fusion reactions and then confined in the plasma. The scattering method discussed above for ion temperature measurements also is suitable for detection of these confined alpha particles. Such experiments are awaited eagerly.

4.2.2 Measurement of Density and Temperature of Neutral and Impurity Species

The presence of neutral species in a plasma affects the propagation of laser light in a variety of ways, such as refraction effects, described in section 2.2.2, non-resonant Rayleigh and Raman scattering effects, and resonant scattering effects such as laser-induced fluorescence (LIF), described in section 2.2.4. Of the methods based on these effects, LIF spectroscopy is the most powerful method for studying neutral and impurity species in high-temperature plasmas.

Tunable wavelength lasers are a necessary part of the apparatus for LIF experiments, because the wavelength of the laser light must be matched to a resonant wavelength in the species to be measured. For most metallic elements, excitation from the ground electronic state is possible, using either the fundamental or the second-harmonic output of a tunable laser. For many gaseous neutral and ionic species, however, such as oxygen, hydrogen and carbon, excitation from the ground state requires laser light at vacuum ultraviolet wavelengths and this light is difficult to generate. Although further development of laser sources is needed before detection of all desired species is possible, recent progress has made possible measurements of some species.

The main species in high-temperature fusion plasmas is hydrogen, in its various isotopic forms. Understanding of fusion plasmas, and in particular particle recycling and particle confinement, requires knowledge of the behaviour of the hydrogen species in the plasma. Information about the mechanisms controlling diffusion of hydrogen to the chamber walls, and adsorption and emission of these species from the wall surfaces, is necessary.

Several excitation and fluorescence schemes for LIF experiments for hydrogen are shown in figure 4.13. LIF methods based on Lyman transitions inherently are difficult because of the short wavelength, but excitation and fluorescence schemes based on the Balmer α transition ($\lambda = 656.3$ nm) are much easier and have been used for many experimental studies. However, because excitation occurs from the excited state, a model such as a collisional-radiative model must be used in order to determine the density in the ground state from these fluorescence measurements. Such measurements have been made on tokamaks, stellarators, and mirror machines, with the density measurements being used to discuss particle confinement and charge exchange mechanisms in these devices. The measurement of hydrogen density in the Compact Helical System (CHS), a heliotron/torsation type of toroidal device at the National Institute for Fusion Science in Japan, is one example of such a measurement. An example of the results of this measurement is shown in figure

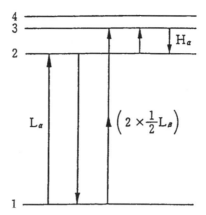

Figure 4.13. Energy level diagram of atomic hydrogen, showing excitation and fluorescence transitions for LIF measurements.

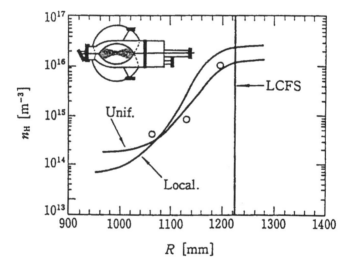

Figure 4.14. Radial profiles of atomic hydrogen density, n_H, measured by LIF spectroscopy using the H_α transition on CHS. The open circles are the measured values and the solid lines show the profile calculated using a simulation. The 'Unif.' result was calculated assuming a hydrogen source that was distributed uniformly around the chamber walls and the 'Local' result was calculated assuming that hydrogen atoms came from the position where the magnetic field lines intersect with the wall. The position marked LCFS indicates the last closed flux surface position.

4.14 [14]. Hydrogen density was measured at different radial positions, and compared with the predictions of simulations based on different sources for neutral hydrogen. From this comparison, the emission processes of hydrogen atoms and molecules from the chamber walls and from dissociation of the

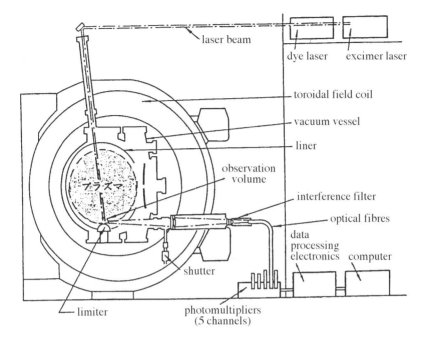

Figure 4.15. Experimental arrangement for measurement of neutral iron atoms in the TEXTOR tokamak. The laser beam was directed towards the iron test limiter, and fluorescence was observed from atoms sputtered from the limiter surface. The laser was an excimer laser-pumped dye laser, operated at 100 Hz repetition rate.

molecules, were discussed, together with the particle confinement times in the plasmas.

The interaction between plasma and chamber walls in high-temperature plasma devices results in species being adsorbed on the wall surfaces and particles from the wall material itself being introduced into the plasma. These impurity species become highly ionized and line emission and bremsstrahlung radiation from these highly ionized particles in the plasma can be a major form of energy loss. Also, impurity species cause large gradients in the radial current distribution, which in turn generate MHD instabilities that disrupt the equilibrium state of the plasma. For these reasons, information about the type of impurity species, their density, spatial distribution, and generation mechanisms is desired in order to learn how to control impurity problems.

Almost all impurity species near the wall are neutral atoms and molecules in their ground states. These species do not emit radiation, because there is little electron impact ionization and excitation due to the low electron temperature and density in this region. LIF spectroscopy has been used to try and measure these species. One example is the measurement of iron atoms sputtered from the test limiter in the TEXTOR tokamak [15]. Figure 4.15 shows the arrangement of

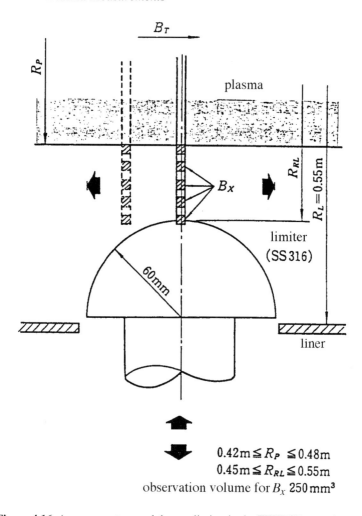

Figure 4.16. Arrangement around the test limiter in the TEXTOR experiment.

the apparatus for this experiment, and the measurement region around the test limiter itself is shown in more detail in figure 4.16. The system is arranged so that the laser probes the scrape-off layer in the region directly above the test limiter. The spectral lines used for excitation and fluorescence are the same as those shown in figure 4.63 (in the following section about measurements in glow discharges). An example of the results is shown in figure 4.17. Part (a) of the figure shows the temporal behaviour of the Fe atom density at one particular position. Part (b) shows the decay length of Fe atoms penetrating into the plasma, determined from a series of measurements at the positions B_x shown in figure 4.16. The set of data measured in this LIF experiment allowed the subject of impurity problems in tokamak plasmas to be discussed quantitatively.

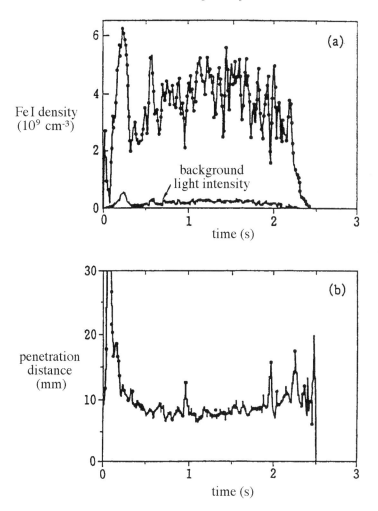

Figure 4.17. Temporal dependence of (a) Fe atom density in the region close to the limiter, and (b) penetration depth of Fe atoms into the plasma.

Experiments using excitation wavelengths greater than about 200 nm are relatively straightforward because tunable dye lasers, discussed in section 3.1.5, can be used to generate these wavelengths. In this wavelength range, a special kind of laser, called the rapid-frequency scan (RAFS) laser, has been developed for measurements of instantaneous spectral profiles. This laser is a dye laser in which the oscillation frequency is scanned over the necessary wavelength interval in a short laser pulse, of about 5 μs length. This laser has been used to measure Doppler profiles, in order to determine the penetration velocity of atoms into the plasma, and Zeeman-split spectral profiles, in order to gain information about magnetic fields [16].

Many other species, including nitrogen, hydrogen and oxygen, require excitation wavelengths of less than 200 nm. A variety of laser sources are being developed for this purpose. As discussed previously, the most important target species is atomic hydrogen and various schemes for excitation from the ground state have been proposed and tested. One of these schemes uses the output beam from an ArF laser, which is frequency shifted by passing the beam though a D_2 Raman cell to generate light at $\lambda = 205$ nm. This scheme, shown in figure 4.13, uses the 205 nm laser light to excite hydrogen atoms from the ground state to the $n = 3$ state (the Lyman β transition) using two-photon excitation. Fluorescence then is observed on the Balmer α wavelength of 656.3 nm. This scheme has the two advantages that vacuum ultraviolet wavelengths are not necessary, and that different wavelengths are used for fluorescence and excitation, which avoids stray light problems. In addition, excitation and fluorescence can be performed along the same optical axis with good spatial resolution. This is possible because the fluorescence intensity, which depends quadratically on the laser power, mainly originates from the region where the laser beam is tightly focused. These advantages mean that this method is the most suitable technique for hydrogen detection in high-temperature plasmas. The first application of this technique was performed on the Heliotron E device at Kyoto University in Japan [17].

4.2.3 Measurement of Electric and Magnetic Fields and Plasma Fluctuations

Electric and magnetic fields and the fluctuations in both these quantities are important properties of high-temperature plasmas. Various laser diagnostic techniques have been proposed to measure these quantities, and some have been developed to the stage of practical use.

Electric field. Measurement of electric field in high-temperature plasmas using electromagnetic radiation generally is difficult. Some methods are based on detection of electric field effects on neutral particle beams that are introduced to the plasma specifically for this purpose. In these methods, emission and/or fluorescence from the introduced particles is measured. However, application of these methods has been troubled by both fundamental and practical difficulties, and none of the methods tested so far has been developed into a reliable practical technique.

Methods do exist for measurement of fluctuations in an electric field, however, and the corresponding fluctuations in electron density. The collective Thomson scattering methods discussed in section 2.2.4 are used widely to detect and to measure the (k,ω) spectra of the electron density fluctuations. In addition, several new methods have been developed more recently for the measurement of electron density fluctuations. These methods include reflectrometry [7], Fraunhofer diffraction methods [18] and laser phase contrast methods [19,20].

Magnetic field. Four methods have been used to measure the magnetic field in

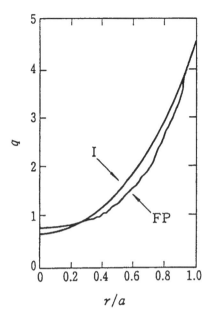

Figure 4.18. Safety factor q determined from polarimetry measurements in the TEXTOR tokamak. The plasma conditions, and I and FP, are the same as for figure 4.3.

high-temperature plasmas. The Faraday rotation method, discussed in section 2.2.2, measures the change in polarization of electromagnetic radiation which is directed through the plasma. Another method measures the polarization angle of electromagnetic waves generated by nonlinear harmonic generation at the upper-hybrid-resonance absorption in the plasma. A third method is based on observing incoherent Thomson scattering at right angles to the magnetic field and measuring the modulation in the spectrum due to oscillations at the electron cyclotron frequency. In addition, laser-induced fluorescence methods can be used to observe Zeeman effects in spectra of particles in the plasma.

The Faraday rotation method is the most developed of these four methods, and it is now a well-established measurement method used on several high-temperature plasma devices. An example of a measurement system is the apparatus installed on the TEXTOR tokamak that is described in section 4.2.1 and shown in figure 4.2 [5]. Using this Faraday rotation measurement system, the distribution of the magnetic field, and from this, the distribution of plasma current, was determined. From the plasma current distribution, the safety factor q for the tokamak operation can be determined. An example of the TEXTOR results is shown in figure 4.18. This result showed that q in the plasma centre had a value of 0.75–0.8, which was an important result because it had been believed for theoretical reasons that it was not possible to generate a stable plasma with $q < 1$.

138 *Plasma Measurements*

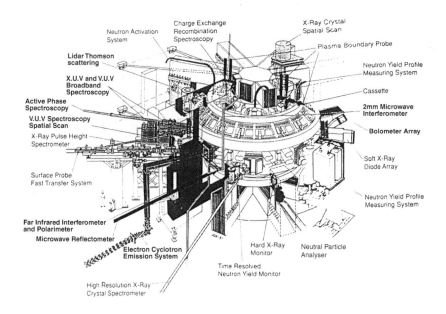

Figure 4.19. Arrangement of diagnostic instruments at the JET tokamak. The instruments discussed in this chapter are shown in bold font.

Most of the measurement methods described in this section are installed in one form or other in modern high-temperature plasma devices. As an example, figure 4.19 shows how diagnostics are arranged on the Joint European Torus (JET) tokamak, which is the world's largest and most successful device for high-temperature plasma research [21]. It can be seen that measurement systems based on the techniques described here, together with many other diagnostic systems, completely surround the torus, and are integrated into the tokamak's structure.

4.3 Laser-Aided Measurements in Discharge Plasmas

Discharge plasmas have many applications that include etching and thin-film deposition for integrated circuit fabrication, surface treatment, welding, cutting, melting, lighting, gas laser excitation, spacecraft propulsion, and plasma display panels. In each of these areas, optimization of the discharge often is performed empirically. In this empirical process, the control parameters of the discharge, such as the electrode shape, the magnitude and frequency of the applied voltage, the gas composition, the gas pressure and the chamber dimensions, are varied in order to optimize the plasma conditions. However, for a systematic understanding of the plasma, it is necessary to view the discharge in terms of the basic physical processes that occur in it. One such approach is to consider the discharge properties to be determined by the following three processes. Initially, an external power source is used to generate a complex electric field in the discharge volume. Electrons absorb energy from this electric field, which determines the electron properties such as density and temperature. The electrons then react with other particles in the discharge to generate the new particle species that are important for that particular application of the discharge. Hence, for any discharge, information about these quantities, the electric field, the electron density and electron temperature, and the reaction products, is necessary for understanding the fundamental mechanisms and subsequent optimization of the discharge. Comparison of experimental results with simulation results might be part of this optimization process.

These three quantities depend very much on the discharge conditions, and have values that range over many orders of magnitude. For this reason, techniques to measure these quantities differ according to the type of discharge and it is difficult to make general remarks about laser-aided diagnostics of discharge plasmas. In this section, some representative examples that describe recent research activities are presented. In this respect, it is to be noted that there has been a great deal of progress in laser-aided techniques for measurement of electric fields and electron properties in glow discharges, and these quantities now can be measured over a wide range of plasma conditions.

4.3.1 Measurement of Electric Field

Electric field measurement techniques have been applied to various kinds of low-pressure glow discharge plasmas, and some of these measurements are described in this section. However, similar laser-aided techniques have not been applied to high-pressure discharges such as arcs. The reason for this is that the best spatial resolution that can be attained by lasers is of the order of the laser wavelength. In low-pressure discharges, where the electric field changes over distances of the order of a millimetre, this resolution is sufficient for accurate measurements. In high-pressure arc discharges, however, the electric field is concentrated within very thin layers, less than 1 µm wide, in the region next to the electrodes. This has prevented the application of laser-aided diagnostics of electric fields in these discharges.

Laser spectroscopic methods for determining the electric field are based on measurement of the Stark effect. Laser spectroscopy for electric field measurement in glow discharge plasmas began with laser optogalvanic (LOG) spectroscopy of helium atoms, reported in 1984 by Lawler *et al* [22]. This group used a tunable laser to excite helium atoms from metastable states to Rydberg states, and measured the excitation spectrum. The magnitude of the electric field was determined from Stark splitting of the Rydberg states. Because Rydberg states have long lifetimes, it is not possible to observe fluorescence from these states, and so the change in plasma impedance caused by the laser excitation, an optogalvanic effect, was monitored. Although excitation from the ground level is preferable, the short wavelength required for this (~ 50 nm) makes direct excitation extremely difficult. However, the metastable states of helium usually have a significant density, and the necessary wavelength for excitation from these states to Rydberg levels is around 300 nm. This radiation can be generated easily using frequency conversion of the output of commercially available tunable dye lasers.

Other electric field measurements continued on from Lawler's original research. In one such study Ganguly et al [23, 24] applied LOG spectroscopy to different discharges. LOG spectroscopy is an accurate method, but is applicable only to DC and low-frequency AC discharges, in which small changes in the plasma impedance can be measured. A further breakthrough came when Greenberg and Hebner extended Lawler's original method to an LIF method, which in principle could be applied to any type of discharge. Although direct fluorescence from the Rydberg states cannot be observed, Greenberg and Hebner succeeded in observing fluorescence from other levels that were populated by collision redistribution of the excited Rydberg atoms [25]. This new method greatly enhanced the potential of the Lawler's method because of the high spatial and temporal resolution that can be achieved with LIF techniques.

The schemes discussed above are based on detection of Stark splitting and shifting of Rydberg states. Gottscho et al [26] used a very different scheme to measure electric fields using Stark mixing of levels in BCl radicals produced in a discharge containing BCl_3. A tunable laser was used to excite a vibrational transition in the BCl radical, and the magnitude of electric field was determined by the ratio of fluorescence intensities of an allowed and a forbidden transition. Fields as small as 4 V mm^{-1} were measured with good spatial resolution.

Electric field measurements using spectroscopy of hydrogen [27], neon [28], CS [29] and NaK [30] also have been reported.

Measurement of Stark splitting and shifting. This first group of electric field measurement methods relies on detection of Stark splitting or Stark shifting of energy levels due to the presence of the electric field. The method based on spectroscopy of helium, which was developed by Lawler [22] and then extended by Greenberg and Hebner [25], is representative of these methods.

In these methods, it is possible to observe spectra using many different excitation transitions. Figure 4.20 shows an example of an excitation and detection scheme. A tunable dye laser is used to excite helium atoms in the discharge on the singlet $2s \rightarrow n = 11$ transition ($\lambda \sim 321$ nm). There are two methods for detecting that excitation. In the laser optogalvanic (LOG) method, the change in plasma impedance caused by the excitation is detected. When the laser radiation is absorbed, atoms are excited to an energy level close to the ionization energy, and subsequent ionization of these atoms affects the electron density and temperature in that part of the discharge. This, in turn, changes the overall impedance of the discharge, which can be measured as a change in the discharge current or voltage. The other method is a laser-induced fluorescence (LIF) technique. In this case, however, fluorescence is not observed directly from the excited Rydberg level because these levels have extremely long

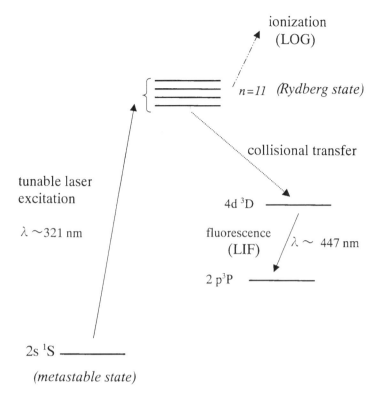

Figure 4.20. An example of excitation and fluorescence transitions in the neutral helium atom that can be used to measure electric field.

radiative lifetimes. However, collisions in the plasma redistribute the excited-state populations, and atoms in Rydberg levels are collisionally de-excited to lower energy states. Fluorescence can be observed from these lower states, and so the excitation can be monitored indirectly. This variation of the standard LIF technique can be called *laser-induced collisional fluorescence* (LICF).

Figure 4.21 shows the theoretical Stark map of the $n = 11$ singlet levels of helium, calculated using quantum theory. Figure 4.22 shows several examples of the theoretical calculation for the Stark shift of the singlet P level for quantum number $n = 7-11$. It can be seen that the Stark shift is much larger for transitions to levels with higher quantum number n and so transitions to higher upper levels are more sensitive to the electric field.

Figure 4.23 shows an experimental arrangement used for electric field measurement by LOG spectroscopy [31]. The apparatus had three main parts: the vacuum/discharge chamber system, the tunable laser and the optogalvanic signal detection circuit. In the experiment discussed here, the plasma was a DC discharge, excited between parallel stainless-steel electrodes with diameter of 40 mm and separation of 10 mm. The applied voltage was in the range of

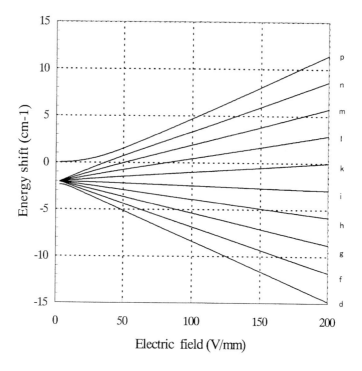

Figure 4.21. An example of the Stark splitting of Rydberg states in a helium atom ($n = 11$).

168–195 V, the discharge current in the range of 1.8–5.7 mA and the gas pressure in the range of 1–3.9 torr.

For the apparatus shown in figure 4.23, the excitation laser source was a tunable dye laser pumped by an XeCl excimer laser. The output from this laser was frequency-doubled using a KDP crystal to generate the $\lambda = 315$–325 nm light that was necessary to excite transitions from the $2s^1S$ metastable level to Rydberg levels with $n = 11$–19. The remaining part of the laser output at the fundamental wavelength was removed using a UV filter. The laser beam was polarized in the direction perpendicular to the electrode surface (i.e. parallel to the electric field) so that the only transitions that were excited were those for which the change in magnetic quantum number Δm was zero. A cylindrical lens was used to focus the light into a sheet-like beam that was parallel to the electrode surface, with dimensions of 0.1 mm × 1 mm. The experiment was performed by scanning the laser wavelength, averaging the LOG signal with a boxcar averager and then recording the spectrum with a pen recorder. The electrodes were mounted on a translation stage so the electrode assembly could be moved with respect to the laser beam without changing the optical alignment.

Transitions between singlet states of helium were used for this study, because the singlet states are more sensitive to the electric field than the triplet

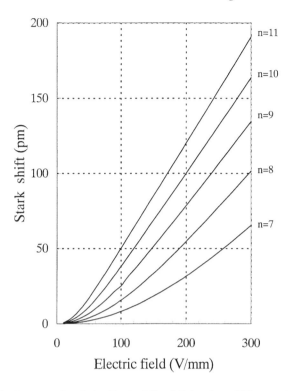

Figure 4.22. An example of the Stark shift of P level at different principal quantum number ($n = 7$–11).

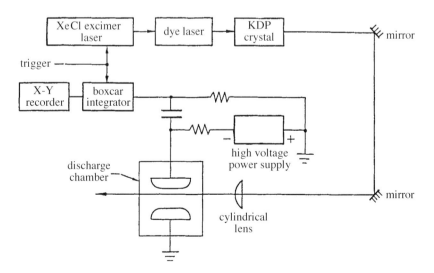

Figure 4.23. Example of an experimental arrangement for an electric field measurement using laser optogalvanic (LOG) spectroscopy.

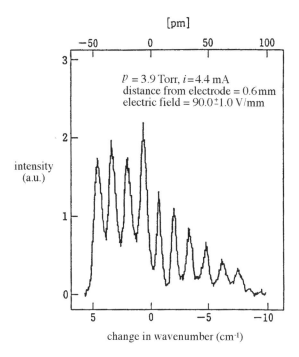

Figure 4.24. Example of a measured spectrum, showing the splitting of the He singlet $n = 11$ levels.

states. Figure 4.24 shows a Stark spectrum measured for the case of $n = 11$ and laser spectral width $\Delta\lambda = 3.2$ pm. For this measurement, the gas pressure was 3.9 Torr, the discharge current was 4.4 mA and the distance from the electrode was 0.6 mm. The electric field was determined for this spectrum to be $E = 90 \pm 1.0$ V mm^{-1}, using the electric field dependence shown in figure 4.21.

Figure 4.25 shows the spatial distribution of electric field in the region close to the electrode, measured in discharges with different discharge current i. The electric field at the surface of the electric field was 80, 140 and 180 V mm^{-1}, depending on i, while it can be seen that the field distribution was linear with distance d for each case.

Figure 4.26 shows the distribution of electric potential, obtained by integrating the electric field distributions shown in figure 4.25 in the region near the electrode. Measurements with a single probe indicated that this plasma was in the anomalous glow discharge regime, and hence nearly all of the potential difference between the electrodes should be in the cathode sheath region. For the cases of the three discharge currents of $i = 1.8$, 4.4 and 5.7 mA, the potentials determined from the electric field measurements, 197 ± 4, 183 ± 5 and 157 ± 5 V respectively, were in good agreement with the total voltage between the electrodes for each case, which were 195, 180 and 168 V.

These measurements were made using LOG spectroscopy of helium atoms.

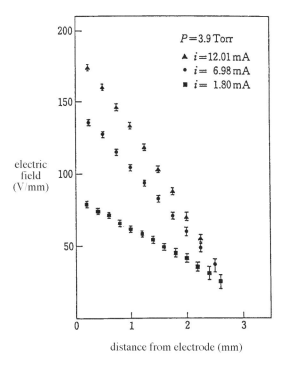

Figure 4.25. Electric field distribution measured in the region close to the cathode of a helium DC discharge.

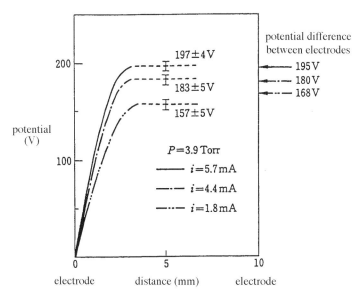

Figure 4.26. Potential distribution in the region close to the cathode of a helium DC discharge, determined from the distributions shown in figure 4.25.

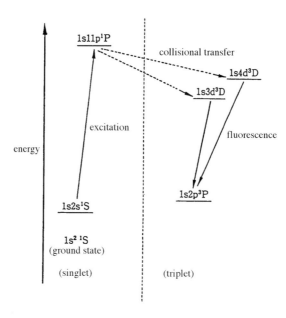

Figure 4.27. Energy level diagram for helium, showing important transitions in the collisional LIF method.

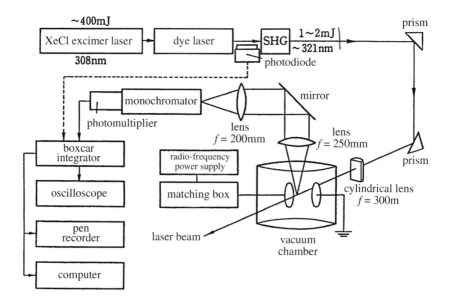

Figure 4.28. Experimental arrangement for an electric field measurement using collisional LIF.

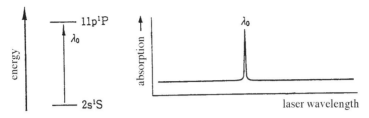

(a) absorption spectrum in the absence of an electric field

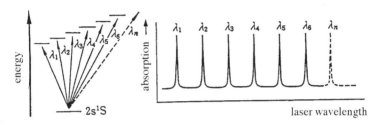

(b) absorption spectrum in the presence of an electric field

Figure 4.29. Examples of collisional LIF signals from electric field measurement.

LICF is a recent extension of this basic technique. The spectroscopic scheme for this experiment, indicating the excitation, collisional and fluorescence steps, is shown in figure 4.27 [32]. In this example, fluorescence was measured on the $1s4d^3D$–$1s2p^3P$ transition, and sufficiently large signals could be measured because the $1s4d^3D$ state has a lifetime of about 40 ns. Figure 4.28 shows an example of an experimental apparatus used for an LICF experiment on an RF plasma. The wavelength of the tunable dye laser was scanned across the wavelength of the $1s2s^1S - 1s11p^1P$ transition. At gas pressure of a few hundred millitorr, the collision frequency is about 2 MHz, and so the lifetime of the excited state, which is the inverse of the collision frequency, is about 500 ns. This is much shorter than its radiative lifetime, which is of the order of microseconds. Hence, the excited atoms are transferred by collisions into other states, including the $1s4d^3D$ and $1s3d^3D$ levels shown in figure 4.27. Fluorescence from these levels can be observed only when there is excitation on the original transition, which will occur only when the wavelength of the dye laser matches that of a transition. Examples of the expected absorption spectra, which also can be considered as fluorescence spectra, are shown in figure 4.29.

In the example discussed here, the plasma was excited by an RF power supply, and so the electric field varied at the frequency of the excitation voltage. However, the fluorescence signal had a decay time of hundreds of nanoseconds, and was detected with a gate width of 300 ns to collect most of the fluorescence signal. A simulation of the spectrum expected from such a measurement, indicated that the measured spectrum corresponded to the maximum electric field present

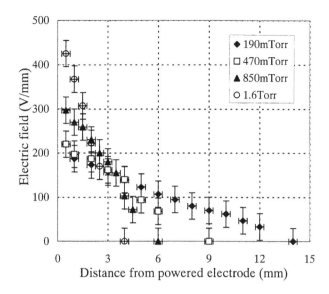

Figure 4.30. Distribution of electric field near the powered RF electrode as a parameter of gas pressure (pressure: 0.19–1.6 Torr, RF power: 100 W, area of electrode: 5×10^3 mm^2).

during the RF cycle. Figure 4.30 shows electric field distributions in the sheath region of the discharge measured in this way for various discharge conditions. The vertical error bars, which represent the accuracy of the electric field, were determined by the accuracy with which the excitation wavelengths could be measured. The horizontal error bars, which represent the accuracy of the spatial position, were determined by the width of the laser beam inside the discharge.

As was done with the results shown in figure 4.26, the absolute reliability of these experiments was checked by comparing the integrated electric field with the voltage that was applied across the electrodes. The voltage drop across the powered electrode sheath was calculated by integrating the electric field distribution shown in figure 4.30, and the comparison is shown in table 4.1. As described above, the measured distribution corresponded to the electric field

Table 4.1. Comparison of the sheath voltages obtained by integrating the electric fields in figure 4.30 with the actual voltages applied to the electrodes.

Pressure (Torr)	Voltage obtained by integrating electric field (V)	Voltage applied to electrodes (V)
0.19	1350 ± 245	1200 ± 20
0.47	1040 ± 250	1100 ± 20
0.85	1000 ± 140	930 ± 20
1.6	950 ± 170	860 ± 20

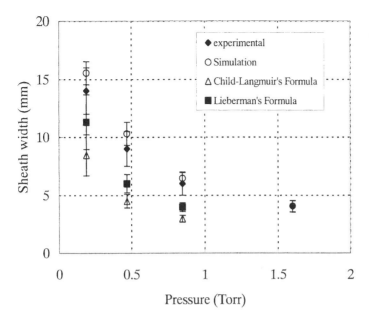

Figure 4.31. Sheath thicknesses obtained from the experiment (♦), compared with the analytical predictions by Lieberman of the maximum RF sheath thickness (■) and the numerical simulation (○). Also shown are the thicknesses obtained from the Child–Langmuir law for a DC discharge (△).

when the sheath was at its maximum extent from the powered electrode (the maximum electric field). From table 4.1, it can be seen that the integrated voltage agreed reasonably well with the voltage that was applied across the discharge. For example, for the condition of $p = 470$ mTorr, the integrated voltage was 1040 ± 250 V, which was consistent with the voltage of 1100 ± 20 V that was applied across the discharge.

The sheath thicknesses determined from figure 4.30 are shown in figure 4.31 [33]. Also shown in the figure are the thicknesses predicted by an analytical theory proposed by Lieberman, the Child–Langmuir theory and a numerical simulation that was performed for this RF plasma. It should be noted that the experimental results, Lieberman's analytical results and the numerical results all correspond to the maximum thickness of the oscillating RF sheath, while the Child–Langmuir results are for a DC sheath.

It can be seen that the sheath thicknesses agree well not only in the general tendency of pressure dependence but also in absolute magnitude, except for those calculated using the Child–Langmuir law. The DC sheath thicknesses calculated using the Child–Langmuir law are about 60% of the other values. This tendency is explained in [34] as being due to the decrease of the space charge by the non-zero time-average electron density in the sheath region. Detailed discussions of these results are available elsewhere [33].

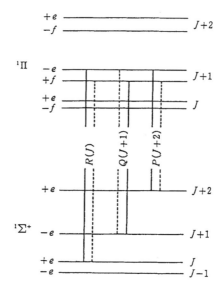

Figure 4.32. Transitions between $^1\Pi - {}^1\Sigma^+$ levels.

LOG methods have an inherent problem of poor spatial resolution because the impedance change that is measured is caused by absorption along the entire laser path in the plasma. Another problem is that the method is restricted to simple discharge types because of the need for low discharge noise. Until recently, both LICF and LOG methods were restricted to discharges containing helium or hydrogen because of the difficulty of calculating Stark effects for other atoms and molecules. However, it has been shown recently that experimental calibration [35] or theoretical calculation [36] can be used to determine electric field effects in other species, and so enable LICF and LOG methods to be used to measure electric fields in discharges containing other gases, most notably argon. It is hoped that similar approaches are used for other gases so that electric field measurements for other plasma conditions become feasible.

Measurement of Stark mixing effects. Stark mixing is the another important Stark effect, in addition to Stark splitting and shifting. In the presence of strong electric fields, energy levels in close proximity can become 'mixed' and transitions that are forbidden according to normal selection rules for electronic transitions, such as $\Delta L = 0, \pm 1$, can become allowed. Because of this effect, emission lines that are usually forbidden can be observed when there is a non-zero electric field. One example of a measurement method that uses this effect is measurement of electric field using diatomic molecules, such as BCl, that have non-zero total orbital angular momentum Λ. When $\Lambda \neq 0$, each rotational energy level is split into doublets, and the energy difference between the two doublet levels is extremely small. An example of this is shown in figure 4.32 [26], which shows part of the energy level diagram for the BCl molecule. The upper energy level band has e and f levels, with the parity of each level being given by $(-1)^J$ and $-(-1)^J$, where J is the rotational quantum number. If, for example, excitation is performed for a P branch transition ($\Delta J = -1$) to an e level when the electric field is zero, then fluorescence can be observed only on the P and R ($\Delta J = +1$) branches. This is because of the parity selection rule, which states that the parity of the particle must change during a transition. However, when there is a finite electric field, the upper e level is mixed with its neighbouring f level, which has opposite parity. This results in fluorescence occurring on the Q branch line ($\Delta J = 0$) as well as the normally allowed P and R branch lines. The ratio of the fluorescence on these lines depends on the strength of the electric field, and this can be used as a method to determine the local electric field.

Fluorescence will occur on all three branches, but if excitation is performed on, for example, the P branch transition, stray light at this wavelength from the exciting laser will make it difficult to accurately measure the fluorescence. In this case, it is more convenient to measure the ratio between the Q and R branch transitions and use this quantity to determine the electric field. Gottscho and co-workers performed a calibration experiment, in which they measured fluorescence from BCl molecules in a constant applied electric field, and obtained the following relationship [37, 38]:

$$\frac{Q}{R} = \frac{\alpha \cdot E^2}{1 + \beta \cdot E^2} + C. \tag{4.3}$$

In this expression, α, β and C are calibration parameters that depended strongly on the gate time over which the fluorescence was measured, the gas pressure, the gas type and the resolution of the measurement spectrometer.

This technique was used in an attempt to measure the electric field in a DC magnetron discharge [39]. The apparatus for this experiment is shown in figure 4.33. The plasma was generated by placing a DC voltage of about 500 V across two parallel, planar electrodes separated by 60 mm. The radius of the charged electrode was 100 mm and the ground electrode had a radius of 50 mm. The magnetron magnetic field was generated by a cylindrically symmetric arrangement of permanent magnets positioned behind the cathode, as shown in figure 4.34. The central post acted as the north pole while the outer ring acted as the south pole. Hence, the magnetic field was approximately parallel to the electrode surface in the region between the poles. This magnetic field resulted in a ring shaped discharge with an average radius of 25–30 mm. Three magnetic fields were used: 0.2 T, 0.4 T and 0.45 T.

The laser system used for this experiment was a XeCl excimer laser with a pumped tunable dye laser. The excimer laser was a Lambda-Physik EMG-203 LMSC laser, with output wavelength of 308 nm, output power of 450 mJ, pulse width of 28 ns and maximum repetition rate of 200 Hz. The dye laser was a Lambda-Physik FL 3002 laser, with an oscillator and two-stage amplifier arrangement. The second-harmonic wavelength was generated using a BBO crystal.

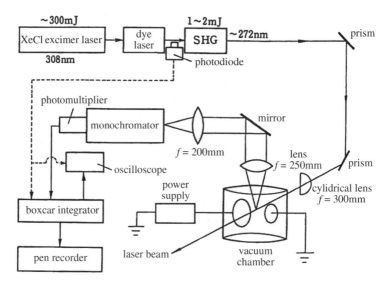

Figure 4.33. Arrangement for measurement of electric fields in a DC magnetron discharge using Stark mixing.

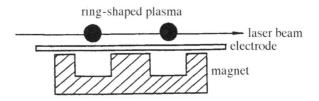

Figure 4.34. Arrangement of a laser beam and the plasma in a magnetron discharge plasma.

Figure 4.35 shows an example of an LIF spectrum obtained in this experiment. The P(12) transition was used for excitation, and the ratio of fluorescence on the Q(11) and R(10) transitions was observed. Figure 4.36 shows this ratio plotted as a function of distance from the cathode. The ratio changes linearly in the region close to the cathode, and then levels out to a plateau value at 3 mm from the electrode. Figure 4.37 shows the electric field distribution determined from this data, calculated using the calibration expression shown in equation (4.3), and appears to indicate a constant electric field at distances larger than about 3 mm from the electrode. This apparent plateau region is believed to be due to the magnetic field influencing the mixing process, and hence, this effect also must be considered for the results at distances less than 3 mm. Ways of including magnetic field effects in the electric field calculation process are now being investigated.

Another method of determining the electric field is based on polarization

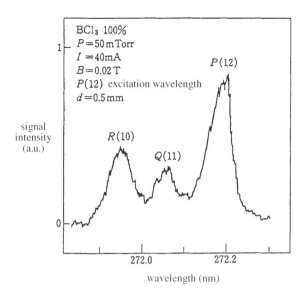

Figure 4.35. Example of an LIF signal measured in the Stark mixing experiment.

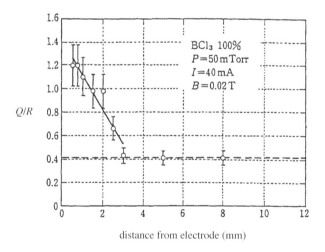

Figure 4.36. Spatial distribution of the measured Q/R intensity ratio.

effects. It is expected that these methods will be applied to measurements in a range of discharges in the near future.

4.3.2 Measurement of Electron Density and Temperature

As for high-temperature plasmas, electron density and temperature in discharge plasmas are determined using laser interferometry and laser Thomson scattering.

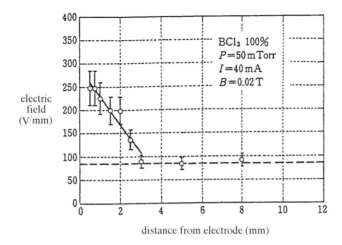

Figure 4.37. Apparent electric field distribution determined without considering magnetic field effects.

Several typical examples are shown in this section. Thomson scattering methods, in particular, have progressed greatly in recent years.

Laser interferometry. A variety of lasers can be used as the laser source for interferometer measurements. If the electron density is relatively low, microwave sources also can be used. These sources are shown in table 2.1. The choice of source depends on the value of $n_e l$ for that discharge, where l is the length of plasma through which the laser propagates and n_e is the electron density. In the past, there have been many measurements of electron density using interferometry. For measurements in arc plasmas in atmosphere, where n_e is of the order of 10^{22-24} m^{-3}, visible lasers are used as the laser source. For measurements in low-pressure glow discharge plasmas, where n_e is 10^{16-18} m^{-3}, far infrared lasers and microwave sources are used.

In this section, interferometry methods are described using the example of an electron density measurement in an impulse discharge that was generated between electrodes in atmosphere. This is a good example of interferometry measurements in discharge plasmas, because there are two different effects that influence the interference fringes. As described in section 2.2.2, fringes are affected both by production of electrons and ions and by the subsequent neutral density changes. Two-wavelength interferometry must be used to distinguish these two effects.

The apparatus for this measurement is shown in figure 4.38 [40]. There are three main parts of the apparatus: the discharge gap system, the two interferometers and the pulse generator that was used to trigger both the discharge gap and the interferometers. The discharge gap system was composed of a pair of sharp-edged tungsten electrodes separated by 5 mm, and a circuit for generating

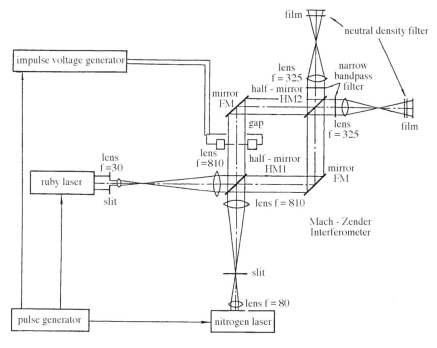

Figure 4.38. Experimental arrangement for simultaneous two-wavelength interferometry measurements.

the discharge. The two interferometers used a ruby laser ($\lambda = 0.6943$ μm, pulse width 5 ns) and a nitrogen laser ($\lambda = 0.3377$ μm, pulse width 5 ns). In the following discussion, the subscripts 1 and 2 refer to these two laser systems.

The arc discharge was generated by applying a voltage of 13 kV to the two electrodes. In order to obtain distinct interference fringes, the optical paths of the beams in both the measurement and the reference arms of the interferometer were adjusted to be approximately equal. The ruby and nitrogen lasers were triggered simultaneously (to within 50 ns), and both laser beams used the same beam splitters, as shown in figure 4.38. Narrowband transmission filters were used at the output of each interferometer to isolate the light at the desired laser wavelength and to reduce the amount of light from the other laser and the background emission from the arc itself. In this way, two separate interference patterns were obtained.

Examples of measured interference patterns from the two interferometers are shown in figure 4.39, for two separate times after the start of the discharge. The photographs show that the fringe shifts are in opposite directions in the central part of the gap and in the regions surrounding the gap. This is because, as explained in section 2, fringe shifts due to electrons are in the opposite direction to those caused by neutral particles.

These photographic images show the line-integrated fringe shifts across the

156 Plasma Measurements

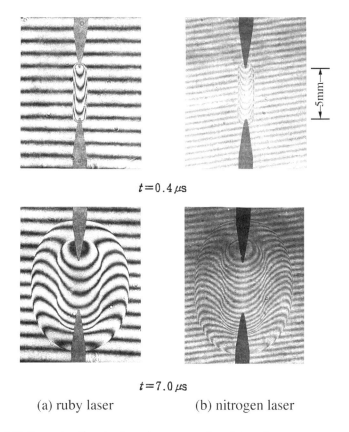

(a) ruby laser (b) nitrogen laser

Figure 4.39. Example of results from the two-wavelength interferometry measurements.

cross-section of the discharge. By using the Abel inversion procedure described in section 2.2.2, it is possible to use these photographic images to calculate the fringe shift per unit length as a function of radius. The theory necessary to interpret the results shown in figure 4.39 is outlined below.

As described in section 2.2.2, the refractive index depends on K_e, K_i and K_n. If the gas density in the absence of the plasma is given by n_0 and the gas density in the plasma is given by n_n, the change in optical path length ΔL of the beam passing through the plasma for a distance L is given by

$$\Delta L = \int_0^L \{K_n (n_n - n_0) + K_i n_i + K_e n_e\} \mathrm{d}l. \tag{4.4}$$

The change in electron density due to the impulse arc discharge will produce a change in the fringe number ΔN in a Mach–Zehnder interferometer of

$$\Delta N = \frac{\Delta L}{\lambda} = \frac{1}{\lambda} \int_0^L \{K_n (n_n - n_0) + K_i n_i + K_e n_e\} \mathrm{d}l. \tag{4.5}$$

By letting the relative refractive indices for radiation at wavelength λ_1 be denoted by K_{n1}, K_{i1} and K_{e1}, the change in fringe number at this wavelength, $(\Delta N)_1$, can be written as

$$(\Delta N)_1 = \frac{1}{\lambda_1} \int_0^L \{K_{n1}(n_n - n_0) + K_{i1}n_i + K_{e1}n_e\} dl. \quad (4.6)$$

Similarly, the change in fringe number for the second interferometer, $(\Delta N)_2$, can be written as

$$(\Delta N)_2 = \frac{1}{\lambda_2} \int_0^L \{K_{n2}(n_n - n_0) + K_{i2}n_i + K_{e2}n_e\} dl. \quad (4.7)$$

For this experiment, we can write $n_e = n_i$, because the ions are singly ionized. In addition, $K_{n1} = K_{n2}$ ($\equiv K_n$) and $K_{i1} = K_{i2}$ ($\equiv K_i$) because K_i and K_n are almost independent of the laser wavelength, as shown in figure 2.12. Using these relationships, the following expressions for n_n and n_e can be derived.

$$n_n(r) = \frac{\lambda_1 K_{e2}\{\Delta N_{u1}(r)\}_1 - \lambda_2 K_{e1}\{\Delta N_{u1}(r)\}_2}{K_n(K_{e2} - K_{e1})} + n_0 - \frac{K_i}{K_n} n_i(r) \quad (4.8)$$

$$n_e(r) = n_i(r) = \frac{\lambda_1 \{\Delta N_{u1}(r)\}_1 - \lambda_2 \{\Delta N_{u1}(r)\}_2}{K_{e1} - K_{e2}}. \quad (4.9)$$

In these equations, $\{\Delta N_{u1}(r)\}_1$ and $\{\Delta N_{u2}(r)\}_2$ are the fringe shifts per unit distance for each wavelength.

These quantities, $\{\Delta N_{u1}(r)\}_1$ and $\{\Delta N_{u2}(r)\}_2$, are shown in figure 4.40. The

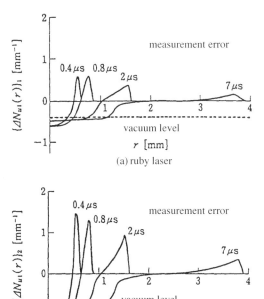

Figure 4.40. Radial distributions of fringe shift per unit distance (after Abel inversion).

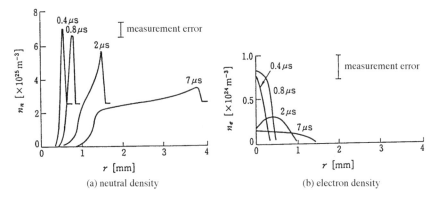

Figure 4.41. Radial profiles of (a) neutral particle density and (b) electron density at different times of an impulse discharge in atmosphere.

vacuum levels shown in the figures, $\{\Delta N_{u1}(r)\}_1 = -0.40$ mm^{-1} and $\{\Delta N_{u2}(r)\}_2 = -0.85$ mm^{-1}, are the fringe shifts that would occur if all the neutral particles present in the gap before the discharge, which was operated at temperature of 15° C and pressure of 1 atm, were removed. Fringe shifts lower than this are observed because of electron density effects. The size of the error bars shown in the figures are mainly decided by uncertainties in determining the fringe shifts from the images.

The neutral density and electron density determined from equations (4.8) and (4.9) are shown in figure 4.41. In order to calculate these results, values of $n_0 = 2.5 \times 10^{25}$ m^{-3}, $K_n = 1.1 \times 10^{29}$ m^3, $K_i = 7.4 \times 10^{-30}$ m^3, $K_{e1} = -2.2 \times 10^{-28}$ m^3 and $K_{e2} = -5.1 \times 10^{-29}$ m^3 were used. One point to note about measurements such as these is that the density of metastable levels of neutral and ionic species have to considered, because these species in metastable states have much larger refractive indices than their ground states. Hence, it is important to verify that the contribution to refractive indices from metastable species can be neglected.

The neutral and electron densities were measured in this way for a variety of different discharge conditions, and the following conclusions were drawn from these measurements. Firstly, the neutral particles inside the arc were virtually completely ionized before the measurements could be made. Secondly, at that time, a pressure of more than 10 atm inside the arc region caused a shock wave with Mach number of 5 to be generated. The velocity (Mach number) rapidly reduced with time. Thirdly, the radius of the region with significant electron density (i.e. the arc channel) increased rapidly for the first microsecond, became constant and then decreased as the current fell away.

This measurement of the spatial and temporal dependence of the density in the arc was the basis of further detailed discussion on many topics, including the temporal change of the arc radius, the average electron drift velocity, average electric field inside the arc, the behaviour of the shock wave and excitation and dissociation in the arc.

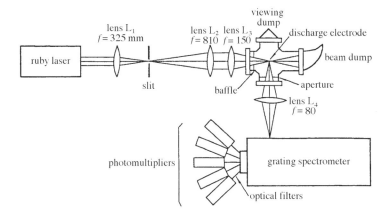

Figure 4.42. Arrangement for a Thomson scattering measurement in an impulse arc discharge in atmosphere.

Thomson scattering. In recent years, Thomson scattering has been used for detailed studies of many different types of arc and glow discharge plasmas. In this section, several examples of these measurements are described.

(i) Thomson scattering measurements in a pulsed arc discharge in atmosphere
The impulse arc discharge discussed above in the section on interferometry also was investigated using Thomson scattering. The apparatus used for this study is shown in figure 4.42 [41]. A pulsed ruby laser with pulse width of 20 ns, spectral width of 0.01 nm and effective pulse energy of 0.2 J was used as the laser source. After passing through the beam shaping optics indicated in the figure by L_1, L_2, L_3 and the slit, the laser beam was focused into the chamber to a spot size of about 0.3 mm. The electrodes were in the centre of the chamber, with the tips of the electrodes being perpendicular to the plane of the figure. Measurements were made in the central plane of the electrodes. Radial profiles of electron density and temperature were measured by moving the lens L_3 along the direction of laser propagation. The axes of the laser and the detection system were perpendicular to each other (i.e. the scattering angle θ was 90°) and both also were perpendicular to the discharge axis. The monochromator had reciprocal dispersion of 0.44 nm mm^{-1}, and a 5-channel optical fibre set, with each channel having width of 47 μm and height of 3 mm, was used to simultaneously transmit the monochromator output to different photomultiplier detectors. The width of the entrance slit of the monochromator was 47 μm, and the width of the total instrument function, including the effect of the laser spectral width, was 2.7×10^{-2} nm. This instrument function is shown in figure 4.43 by the dashed line. The relative sensitivities of the five photomultipliers used to detect the scattered signal was calibrated by measuring the Rayleigh scattered signal from nitrogen gas.

An example of a measured scattered spectrum is shown in figure 4.43. The

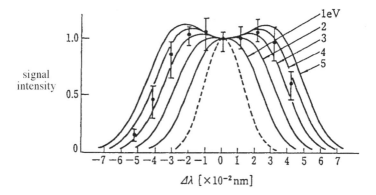

Figure 4.43. Example of a measured Thomson scattered spectrum, for $t = 0.4$ μs and $r = 0$. The dashed curve is the instrumental function, while the solid lines are theoretical spectra, calculated for the temperatures shown.

central wavelength ($\Delta\lambda = 0$) corresponds to the ruby laser wavelength of 694.3 nm. The measurement was made at $t = 0.4$ μs, and on the discharge axis. The data shown in the figure are the average of signals from seven laser shots and the error bars represent the scatter of those measurements.

The measurements made in this experiment were in the regime of $k\lambda_D < 1$: the regime of collective Thomson scattering. Hence the scattering intensity and the spectrum depended on the electron density n_e, the effective charge $\overline{Z}$ and the plasma temperature T. For this arc plasma in atmosphere, $T = T_e = T_i$ could be assumed because the particle collision frequency was so high that thermodynamic equilibrium was well established. In order to determine T and n_e from the scattering data, the data were fitted with a curve based on a combination of Thomson scattering theory and the Saha equilibrium equation, using the following iterative procedure. Initially, $\overline{Z}$ was chosen to be 1, and values of T and n_e were determined from the curve that best fitted the measured data. The pressure then was calculated, using $p = kT(n_e + n_i)$, and the particle densities, including the ion density n_i for each charged state, were calculated using the Saha equilibrium equation with the previously determined p and T. These values of n_e, T and $\overline{Z}$ then were used to calculate a theoretical Thomson spectrum that could be compared with the measured spectrum. This process was repeated with different values of $\overline{Z}$ until values of T and n_e that best fitted the measured data were obtained. For the data shown in figure 4.43, values of $\overline{Z} = 1.5$, $T = 3.4 \pm 0.5$ eV and $n_e = (1.25 \pm 0.10) \times 10^{25}$ m^{-3} were determined using this procedure. Theoretical spectra calculated using these values of $\overline{Z}$ and n_e but different values of T, ranging from 1 eV to 5 eV, also are shown in the figure.

Radial profiles of T and n_e for this arc discharge are shown in figure 4.44. Profiles for four different times in the arc discharge are shown: $t = 0.4$ μs, 0.8 μs, 2 μs and 7 μs. The points marked by solid circles are the electron densities while those marked by solid triangles are the neutral particle densities, which were

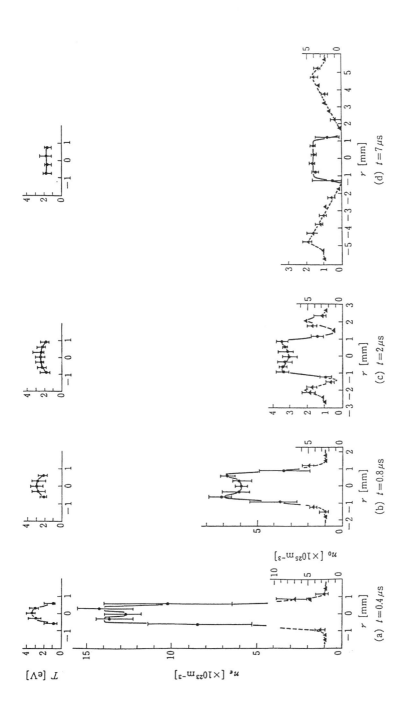

Figure 4.44. Measurements of spatial distributions of temperature T, electron density n_e (black circles) and neutral density n_n (black triangles), inside an impulse arc discharge in atmosphere.

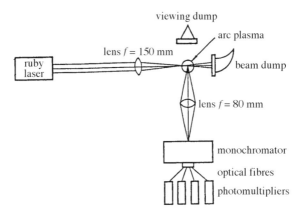

Figure 4.45. Arrangement for a Thomson scattering measurement in a DC arc discharge in atmosphere.

determined from measurement of Rayleigh scattering signals. The distribution of n_n determined from these measurements is indicated by the dotted lines. In the data for times of 2 µs and later, it can be seen that there is a peak of n_n away from the centre. The neutral density then falls away to atmospheric values at larger radii. This peak corresponds to the position of a shock front. At earlier times in the arc, the main part of the arc and the shock wave cannot be distinguished, because the spatial resolution of the measurement (0.3 mm) was not sufficient for this.

The results show that both the electron density and the plasma temperature fell sharply within the first microsecond, and then changed more gradually. This agrees very well with the results obtained in the interferometry measurement, shown in figure 4.41. The electron density profile measured with laser scattering contains peaks away from the centre of the arc. These could not be seen in the profiles obtained using interferometry in the early stages of the arc (at $t = 0.4$ µs and $t = 0.8$ µs), due to the poorer spatial resolution of that instrument. These peaks, however, can be seen at $t = 2$ µs.

Measurements similar to these were performed for a variety of experimental conditions, providing information about arc properties such as the thermodynamic quantities, and the propagation and dissipation of the shock wave.

(ii) Thomson scattering measurements in a DC arc discharge in atmosphere

The experimental arrangement for measurement of properties of a DC arc discharge using Thomson scattering is shown in figure 4.45 [42]. The laser source was the same ruby laser used in the Thomson scattering measurement of the pulsed arc discharge described above, with pulse width of 20 ns and output energy of 1 J. The laser beam was focused by a lens with focal length 150 mm, to a beam spot with diameter of about 1 mm. The arc had a diameter of about 10 mm, so this beam diameter provided sufficient spatial resolution for measurements within the arc. The scattering angle was 90°. An $f = 80$ mm lens was used to collect the scattered light, which then was dispersed using a monochromator with reciprocal dispersion of 6 nm mm^{-1}. As shown in the figure, a four-channel fibre/detector system was used to detect the scattered spectrum. The fibre bundles each had height of 3 mm and width of 47 μm. This system enabled the spectrum to be measured in a single laser shot. The instrument function of the monochromator had a full width at half maximum value of 0.42 nm. This width was sufficiently narrow so that the Thomson spectrum could be resolved with reasonable accuracy, while also enabling sufficiently large signals to be measured at each spectral point. The relative sensitivity of each fibre/photomultiplier set was calibrated using a standard lamp.

The electrodes used to generate the arc were made of carbon rods coated with copper, and were 7 mm in diameter. The arc was set in the direction perpendicular to the plane of figure 4.45. Measurements were made in the central plane between the electrodes. The electrodes were on a translation stage that could be moved perpendicularly to the direction of the laser beam. This was used to move the discharge with respect to the fixed laser beam and detection system so that radial profile measurements could be made. For absolute calibration of the detection system, the electrode apparatus was replaced with a chamber that could be evacuated. The Rayleigh scattered signal from nitrogen gas at pressure of 1 atm was measured, using the same ruby laser and detection system as for the Thomson scattering experiment. For this calibration, the transmission of the window used to introduce the laser into the chamber was determined by measuring the transmitted laser light intensity, and the transmission of the observation window was measured using a standard lamp.

Figure 4.46 shows examples of Thomson scattered spectra measured on the axis of the arc for different values of the arc current. The horizontal axis is the wavelength difference of the measured wavelength from that of the ruby laser (λ = 694.3 nm). The vertical axis is the output voltage of the photomultiplier. Each point is the average of signals from 10 laser shots, and the error bars represent the scatter of the measured values around the average value. The solid lines in the figure are theoretical spectra, calculated using equation (2.31) for different values of temperatures. For the theoretical calculation, the pressure across the arc column was assumed to be 1 atm. It can be seen that the calculated and measured spectra agreed well, and that the temperature could be determined with an accuracy of about ±300 K.

For the data shown in the figure, measured at currents of 5 A, 10 A, 15 A and 20 A, the temperatures were determined to be 7000 K, 7500 K, 7800 K and 8000 K respectively. The electron densities determined from the Saha equilibrium equation were 5.7×10^{20} m^{-3}, 1.2×10^{21} m^{-3}, 1.7×10^{21} m^{-3} and 2.2×10^{21} m^{-3}, respectively. These density values agree well with measurements made for the same conditions using an interferometer.

The temperature profile inside the arc was measured by moving the measurement point inside the arc, using the translation apparatus mentioned previously. The results of these measurements are shown in Figure 4.47. For all four cases shown, the temperature distribution was reasonably flat from the centre

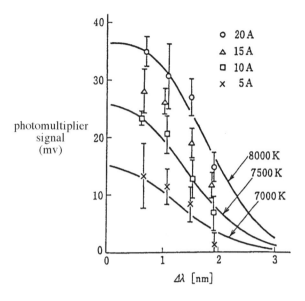

Figure 4.46. Thomson spectra measured for different discharge currents in a DC arc discharge.

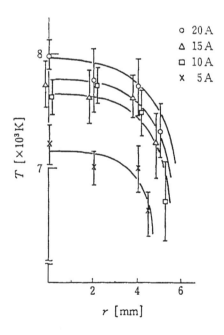

Figure 4.47. Spatial distributions of electron temperature inside a DC arc discharge.

to a radius of about 4 mm, and then decreased rapidly with increasing radius.

These techniques allowed the temperature to be determined with accuracy of about 5%. This accuracy is much greater than was obtained in previous measurements that used passive techniques such as the plasma emission method that was discussed earlier and shown in figure 4.1 [3].

(iii)   Thomson scattering measurements in an discharge-pumped excimer laser
Excimer lasers can be divided into two categories: discharge-pumped lasers and electron-beam-pumped lasers. Lasers pumped with electron beams can be used to produce large energy laser pulses and/or short wavelength light, but are not used widely because the apparatus is large and high repetition rates cannot be obtained. Discharge-pumped lasers, however, can be used at high repetition rates, which means that high average output power can be achieved. In addition, compact, efficient systems can be made and the laser apparatus is reasonably simple and reliable. These are all desirable features for a practical laser. Recently, discharge pumped excimer lasers have been used as light sources for the lithography step of semiconductor fabrication.

Optimization of the discharge and greater understanding of the excimer formation process is required for further improvement of the properties of discharge-pumped excimer lasers. Information about the relationship between the excimer density and the discharge conditions is desirable. The process by which excimers are formed is based on electrons in the discharge volume absorbing the power that is applied via the electrode voltage, and then colliding with other particles to generate new particle species. Hence, information about electron properties such as the electron density and velocity distribution functions is necessary to fully understand the excimer formation process.

In an experiment aimed at obtaining this electron information, Thomson scattering was used to measure electron density n_e and the electron velocity distribution function $f(v_e)$ in an excimer laser discharge [43,44]. These measurements were combined with the results of simulations of the discharge process in order to gain an understanding of the excimer formation process.

There were several difficulties in this experiment that had to be considered before the apparatus could be designed. Firstly, it was expected that the electron density would be of the order of 10^{21} m^{-3}, much smaller than the background gas density, which is $\sim 10^{26}$ m^{-3} at the pressures of several atmospheres that are typical for excimer lasers. Hence, Rayleigh scattering from the gas particles would be extremely large compared with the Thomson scattering from the electrons. Secondly, the discharge occurs over a time of 30–40 ns, and the plasma properties vary rapidly during this period. Hence, temporal resolution of about 1 ns was desirable.

To overcome the first problem, a double monochromator with high rejection efficiency was used. The Rayleigh scattered light appears only at the laser wavelength whereas the Thomson scattered spectrum is broadened by the Doppler motion of the electrons. Use of the double monochromator reduced the effect of the Rayleigh signal on the measurement of the Thomson spectrum. The

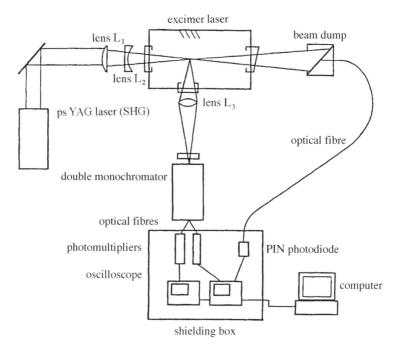

Figure 4.48. Arrangement for a Thomson scattering measurement in the plasma of a discharge-pumped excimer laser.

second problem was dealt with by first using a YAG laser with pulse width of 10 ns to obtain rough information about the electron properties, and then using a different YAG laser with sub-nanosecond pulse width to measure with good temporal resolution.

The apparatus used for the experiment that used the sub-nanosecond laser is shown in figure 4.48. The excimer laser discharge studied in this experiment had a chamber with length 800 mm, width 400 mm and height 560 mm. The distance between the electrodes was 20 mm. The discharge was initiated using pre-ionization discharges generated by a set of 59 pre-ionization pins located along the length of the discharge region. For observation of scattered light, a single pin was removed from the side of the electrode system, so that the scattered light could be observed through a window in the side of the discharge chamber.

The beam from the YAG laser was manipulated with two cylindrical lenses so that the beam, which passed along the discharge axis, had minimum size at the measuring point. The scattered light was collected perpendicularly to the laser beam direction by a lens L_3 that had focal length of 300 mm. A viewing dump and light baffles were used to reduce the amount of stray laser light entering the detection system. The scattered light was dispersed by the double monochromator, whose output was sent via optical fibre bundles to a

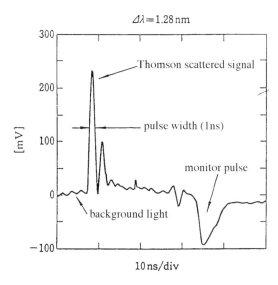

Figure 4.49. Example of a Thomson scattered signal from a discharge-pumped excimer laser. $\Delta\lambda$ is the wavelength separation from the measured wavelength and the laser wavelength.

photomultiplier. The output of the photomultiplier was displayed on a digital oscilloscope and then transferred to a computer.

The sub-nanosecond laser source was a Q-switched YAG laser with active mode-locking and two amplifier stages. A second-harmonic-generation crystal was used to generate light at 532 nm for the scattering experiment. The laser output had pulse width of 0.3 ns and energy of 300 mJ.

The double monochromator was a JASCO CT-25CD model, with focal length of 250 mm. The grating was mounted in the Czerny–Turner arrangement, which minimizes the stray light. The double monochromator consisted of two single monochromators arranged in series, with the exit slit of the first monochromator acting as the entrance slit of the second monochromator. In this arrangement, the theoretical rejection is about 10^{-9}, where rejection is defined quantitatively as the ratio of the transmitted light intensity at any wavelength other than the set wavelength to the transmitted light intensity at the set wavelength. Although the actual rejection in an experiment is higher than this theoretical value, the rejection of the double monochromator is extremely good compared to the rejection of 10^{-4} of most single-stage devices. Hence, this spectrometer was ideally suited to this scattering experiment.

The temporal response of the photomultiplier was very important because the Thomson scattering signal generated by the YAG laser also was in the form of a short pulse. For this reason, a photomultiplier incorporating a microchannel plate (Hamamatsu R2024U), with response of 0.3 ns, was used. The oscilloscope used to view and digitize the signal was a Sony-Textronix MP1101 model, with temporal response of 0.53 ns.

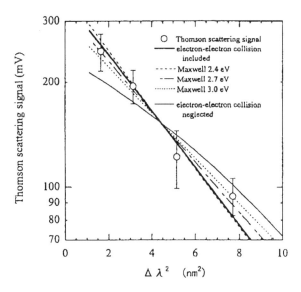

Figure 4.50. Thomson scattering spectra for a gas mixture of Kr/F$_2$/Ne = 30 Torr/1.5 Torr/3 atm.

An example of a measured signal is shown in figure 4.49. For this measurement, the laser was operated in a gas mixture of Kr/F/Ne with pressures of 30 torr/1.5 torr/3 atm, which is a standard gas mix for a KrF laser. The relative timing of the measurement (YAG) laser and the excimer laser was monitored by detecting part of the laser beam using a photodiode. This signal was passed through a pulse delay circuit and then combined with the scattered light pulse from the photomultiplier. This timing signal can be seen clearly in figure 4.49.

Figure 4.50 is an example of a measured spectrum, showing the intensity of scattered signals measured at different wavelengths. The measured spectral points could be fitted with a straight line (on this graph with axes of $(\Delta\lambda)^2$ and log(signal)), and this indicated that $f(v_e)$ was a Maxwellian distribution. A numerical simulation of the discharge for these conditions predicted that $f(v_e)$ would be a non-Maxwellian function. This result is shown by the thin line in figure 4.50. The measurements, however, indicated that $f(v_e)$ was, in fact, a Maxwellian distribution, and so the model used for the simulation was altered to take into account electron-electron collisions, which had been neglected previously. The results of the newer simulation are shown in figure 4.50 by the thick solid line, and agree satisfactorily with the measurements.

From the spectrum shown in figure 4.50, T_e was determined to be 2.7 ± 0.4 eV, and from the total Thomson scattered signal intensity, the electron density was determined to be $n_e = (1.8 \pm 0.2) \times 10^{21}$ m^{-3}. Electron density and temperature were measured in this way at different times throughout the pulse of the excimer laser discharge. An example of these results is shown in figure

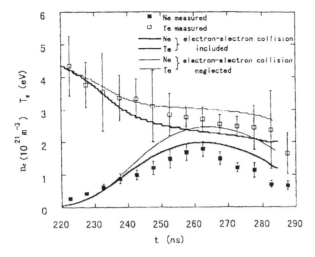

Figure 4.51. Temporal evolutions of electron temperature T_e and electron density n_e for a gas mixture of Kr/F$_2$/Ne = 30 Torr/1.5 Torr/3 atm.

4.51 together with the results of simulations of the discharge. The time shown on the horizontal axis represents the time from the trigger of the excimer laser's thyratron switch. The plotted points are the average of the data from ten laser shots and the error bars were determined by the scatter of the data. The results of the two simulations, one that neglected electron–electron collisions and one that included these collisions, also are shown. The simulation that neglected electron–electron collisions predicted that $f(v_e)$ was non-Maxwellian, and so the results for this case show the average electron energy, rather than the electron temperature. From figure 4.51, it is clear that the measured electron density agreed much better with the results of the simulation that included electron collision. The electron temperatures predicted by both simulations are within the experimental errors.

Measurements such as these were performed for different gas mixtures and different gas pressures. These results allowed the simulation of the discharge to be improved by comparing its predictions with experimental results. This led to better quantitative understanding of phenomena occurring in discharge-pumped excimer lasers.

(iv)   Thomson scattering measurements in an ECR glow discharge

Technology used for fabrication of microelectronic devices has improved continually and rapidly for many years. Much of the progress in this area, in the development of improved plasma sources for etching applications and for plasma chemical-vapour deposition (plasma-CVD), has been achieved on an empirical rather than a systematic basis. However, a combination of empirically based knowledge and greater understanding of the processes occurring in these plasmas is needed for further progress in this field. A necessary part of this

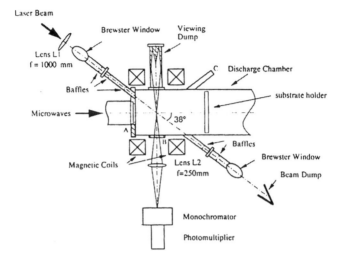

Figure 4.52. Arrangement of an ECR chamber configured for a Thomson scattering experiment. The scattering angle was 52°. Ports A and C were used for the scattering configurations shown in figure 4.55.

research is quantitative information about basic plasma parameters such as electron density and temperature.

The types of glow discharges used for plasma processing applications have relatively low electron densities, and until recently it was difficult to apply Thomson scattering to these plasma sources. However, the use of high repetition rate lasers and data accumulation techniques has enabled Thomson scattering to be used for accurate measurement of electron properties even in these low-density plasmas. To illustrate this progress, this section contains a description of measurements made in an electron cyclotron resonance (ECR) plasma, and the following section shows results obtained from an inductively coupled plasma (ICP).

An ECR plasma is a glow discharge plasma that is generated by absorption of microwaves in a chamber that has a relatively strong applied magnetic field. Electrons resonantly absorb the microwave power at places in the chamber where the microwave frequency ω matches the electron cyclotron frequency $\omega_c = eB/m_e$. Usually, 2.45 GHz microwaves are used, and so the magnetic field corresponding to the resonance condition is $B = 87.5$ mT. The efficient electron heating method means that a plasma with relatively high electron density can be generated at low gas pressure. All the data discussed here were measured in discharges operated in argon gas.

In the first experiment described here, an anisotropy in the electron temperature perpendicular and parallel to the applied magnetic field was observed. Figure 4.52 shows the apparatus used for this experiment [45,46]. The light source for the scattering experiment was a YAG laser operated at the

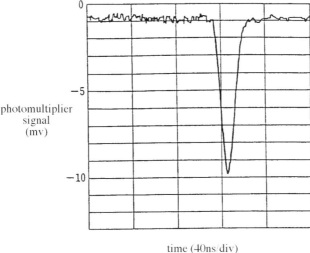

Figure 4.53. Example of a raw Thomson scattered signal. The signal was measured at the centre of a mirror-shaped magnetic field and at $\Delta\lambda = 2$ nm, in a plasma with pressure of 1 mTorr and input microwave power of 570 W.

second-harmonic wavelength of 532 nm. The laser had pulse energy of 500 mJ, pulse width of 10 ns, repetition rate of 10 Hz and beam diameter at the laser of about 10 mm. The scattered light was collected by a lens with focal length of 240 mm. The spectrometer had focal length of 250 mm and reciprocal dispersion of 6 nm mm^{-1}, and the photomultiplier was an R1333 model made by Hamamatsu Photonics. The scattered signal was averaged on a digital oscilloscope and the transferred to a computer for further analysis.

An example of a Thomson scattered signal is shown in figure 4.53. The signal is the average of 1024 laser shots. In this example, the signal-to-noise ratio, determined by the size of the signal against fluctuations in the background light, was more than 30. Figure 4.54 shows an example of a spectrum measured by detecting Thomson scattered signals at five different wavelengths. The electron temperature determined from this data, using equation (2.31), was $T_e = 3.35 \pm 0.19$ eV. The electron density, determined from a Rayleigh scattering calibration of the signal intensity, was $n_e = (2.4 \pm 0.2) \times 10^{17}$ m^{-3}.

Figure 4.55 shows the scattering configurations used to measure the electron perpendicular and parallel to the magnetic field. The scattering vector diagram for each case is shown explicitly. Figure 4.56 shows n_e and T_e as a function of the gas pressure. The density rose continuously with increasing pressure, before approaching a constant value at pressures higher than about 1 mTorr. The electron temperature was measured both perpendicular and parallel to the magnetic field, with the apparatus shown in figure 4.55. The results show that the electron temperature began to increase as the pressure was reduced

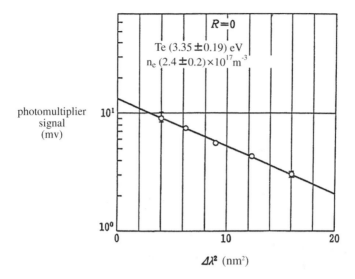

Figure 4.54. Example of a measured Thomson spectrum, obtained for the same plasma conditions as for the signal shown in figure 4.53.

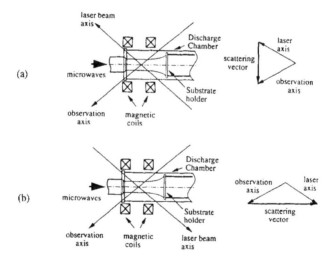

Figure 4.55. Experimental configuration for observing scattering from the electron velocity distribution function, (a) perpendicular to the magnetic field and (b) parallel to the field.

below 1 mTorr. The results also show that there is a temperature anisotropy at low pressures. This kind of observation would not have been possible without the use of Thomson scattering. The appearance of this anisotropy at low pressures can be attributed to a combination of two factors: energy absorption by the electrons occurring mainly in the direction perpendicular to the magnetic

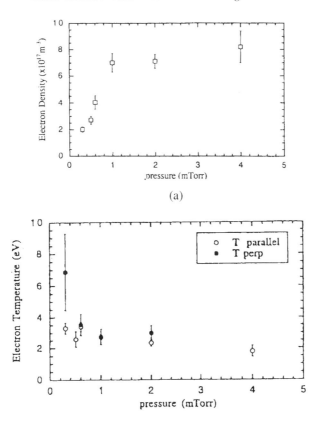

Figure 4.56. Pressure dependence of (*a*) electron density n_e, and (*b*) electron temperature T_e parallel and perpendicular to the magnetic field. The microwave input power was 500 W.

field due to electron cyclotron resonance heating, and the lack of thermalizing electron collisions in these lower density, lower pressures conditions.

In the next experiment described here, the spatial structure of the ECR plasma was studied. Figure 4.57 shows a diagram of the ECR chamber [47]. Observation at the ECR position (referred to as the source region) and the substrate position (downstream region) was possible.

Figure 4.58 shows the distributions of temperature and density that were measured in these two regions from a discharge with argon gas pressure of 1 mTorr and microwave power of 750 W. These distributions depend on various parameters in addition to gas pressure and microwave power, such as the chamber shape and the magnetic field configuration. For this particular case, it can be seen that in the source region there is an off-axis peak in electron temperature and a steep density gradient in electron density. These features, though present, are less distinct in the downstream region. Also shown in the figure are the results of a simulation in which the spatial distribution of

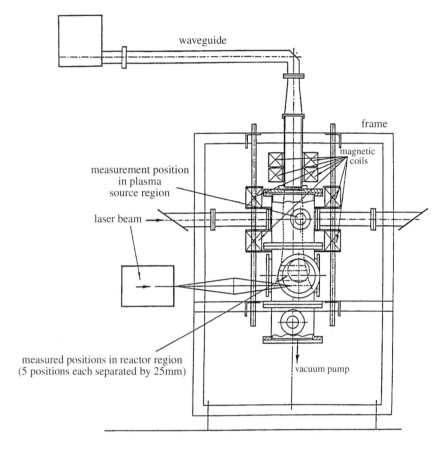

Figure 4.57. Diagram of the ECR discharge chamber used for Thomson scattering experiments, showing the laser entrance and exit window, and the observation windows.

microwave power deposition was assumed. Comparison of the results indicates that the features observed in the distributions of electron temperature and density were reproduced well by the simulation.

In this experiment, measurements were made as functions of many different parameters, including gas pressure, input power, magnetic field configuration and the position of the ECR resonance position. The results of these studies were used to quantitatively discuss the mechanisms that determine internal plasma structures such as the radial and axial potential distributions.

(v) Thomson scattering measurements in an ICP glow discharge

An inductively coupled plasma (ICP) is a discharge that is generated by applying radio frequency (RF) power to coils located either inside or outside the discharge chamber. Relatively high-density plasma can be produced at low gas pressure without the need for an externally applied magnetic field. For this reason, ICP discharges recently have been used for plasma processing

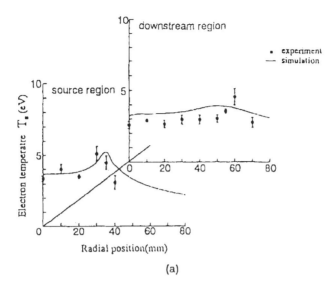

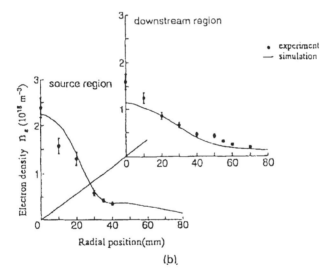

Figure 4.58. Radial profiles of (*a*) T_e and (*b*) n_e in the source and downstream regions of the ECR discharge. The plotted points are experimental values and the full curve is a prediction of a simulation conducted for the same conditions.

applications such as etching. There are several possible coil arrangements for ICP discharges. The one used in the study described here is shown in figure 4.59 [48,49]. In this discharge, a planar three-turn spiral coil was used to supply RF power of 500 W at 13.56 MHz to the plasma. The laser and detection systems used for the Thomson scattering measurements were similar to those used for

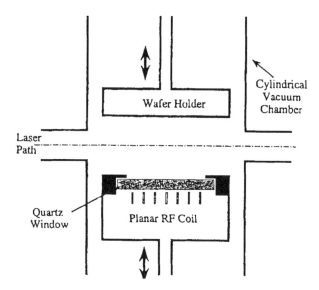

Figure 4.59. Schematic diagram of the ICP discharge. The height and diameter of the chamber were 1000 mm and 300 mm, respectively. The coil chamber and the upper plate could be moved so that measurements could be made at different axial positions without moving the laser axis.

the ECR plasma studies discussed above, and so will not be described here. The data discussed here were measured in discharges operated in argon gas.

The first results discussed here concern the measurement of electron energy distribution functions (EEDF) in the ICP discharge. Figure 4.60 shows Thomson scattering spectra that were measured at different pressures for a fixed RF input power of 500 W. As mentioned previously, if the spectrum can be fitted by a straight line when plotted on these axes (($\Delta\lambda)^2$ vs log(intensity)), then the EEDF is a Maxwellian distribution. It can be seen that the spectra measured at higher pressure correspond to Maxwellian EEDFs, but the EEDFs became progressively less Maxwellian as the pressure was decreased. Similar measurements made at different input powers and pressures indicated that the EEDFs deviated from Maxwellian distributions at low input powers and/or low gas pressures.

The reason this can be understood by considering the competition between the process by which electrons absorb energy, and the process by which they diffuse in velocity space. The characteristic time for the EEDF to approach a Maxwellian distribution is given by the following self-collision time t_{ce} [50],

$$t_{ce} = 3.8 \times 10^{11} \frac{(T_e)^{3/2}}{n_e \ln \Lambda} \text{ s} \qquad (4.10)$$

where $\ln \Lambda$ is the coulomb logarithm, and T_e and n_e are in units of eV and m^{-3}, respectively. t_{ce} is related to the electron–electron collision frequency ν_{ee} by

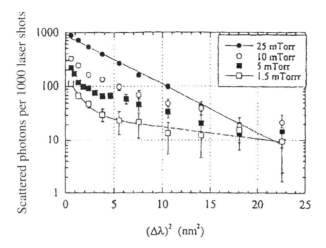

Figure 4.60. Spectra measured for different pressures in the ICP discharge. A straight line fit to the data indicates that the spectrum is Gaussian.

$t_{ce} = 1.1 \times (1/\nu_{ee})$. It can be shown that the EEDF deviated from a Maxwellian distribution for conditions when the electron heating time, defined as the time for electrons to gain energy from the inductive electric field, is sufficiently small compared with the t_{ce}.

For analysis purposes, the EEDFs measured for conditions below 10 mTorr were characterized by two electron temperatures, a high temperature T_1 and a low temperature T_2. When the RF power was switched off, it was found that T_1 approached T_2 within a few μs. This can be explained in terms of the energy exchange time $t_E = t_{ce}/1.1$ [50]. For a typical case of $T_e = 7$ eV and $n_e = 3 \times 10^{17}$ m^{-3}, t_E can be calculated from equation (4.10) to be 2.2 μs. This is in good agreement with the measured value.

The dependence of the electron properties on the discharge parameters also were measured. Figure 4.61 shows the pressure dependence of the electron temperature and density, and the neutral density, measured in the centre of the discharge at a fixed power of 500 W. For this experiment, the aspect ratio of the discharge, defined as the ratio between the axial length and the radius of the plasma, was 2. The results show that electron temperature increased gradually as pressure was decreased in the range 20 mTorr to 5 mTorr, and then increased more sharply at lower pressures. The electron density decreased linearly with decreasing pressure. Figure 4.61(c) shows the neutral particle density in the plasma, determined from Rayleigh scattering measurements. The dotted line indicates the neutral density in the chamber when the RF power was turned off. The Rayleigh scattering cross-sections for atoms in metastable states are several orders of magnitude larger than that for the ground state atoms, and so the contribution of scattering from metastable atoms has to be considered in these measurements. In this case, this contribution was confirmed to be negligible. The decrease of neutral

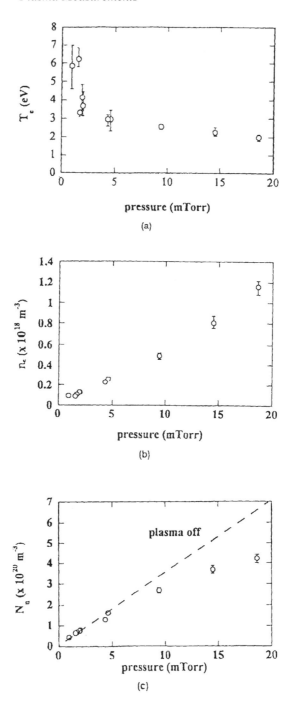

Figure 4.61. Pressure dependence of (*a*) electron temperature T_e, (*b*) electron density n_e and (*c*) neutral particle density N_n.

density in the plasma, by as much as 40%, could be attributed to the neutrals gaining energy from neutral-ion collisions and leaving the plasma centre.

The results shown in figure 4.61, as well as other results measured at different discharge conditions, could be explained using a simple model of the discharge based on a zero-dimensional energy balance equation.

4.3.3 Measurement of Reaction Products

Methods such as laser interferometry and Hook interferometry can be used to measure properties of neutral particles and ions if the density of these species is sufficiently high. In glow discharges, however, and particularly for low-pressure conditions, most reaction products exist in very small densities, and alternative techniques are necessary. Laser spectroscopic methods such as laser absorption and laser-induced fluorescence are widely used for this purpose. CARS (coherent anti-stokes Raman scattering) and DFWM (degenerate four-wave mixing) also have been used.

There are many examples that could be used to illustrate these techniques. In this section, the example of LIF measurements of species ejected from the target electrode in a sputtering discharge is described. These discharges are used widely for deposition of certain types of thin films, and the density and velocity of sputtered particles are important properties that influence thin-film quality.

Deposition of thin films on substrates of various materials is an important step in the fabrication of most microelectronic devices. In the past, vacuum evaporation techniques were widely used, but this technique is not always suitable for many recent applications, such as USLI and liquid crystal display manufacture. Sputter deposition has become one of the standard techniques for deposition of thin films in applications where vacuum evaporation cannot be used.

In sputter deposition, ions from the plasma are accelerated by the sheath electric field and strike an electrode made from the material that is to be deposited. The ions strike the electrode surface with high energy and cause particles of the target material to be ejected from the electrode surface. In the experiment described here, the velocity of the sputtered particles was measured directly with an LIF technique [51].

The experimental apparatus is shown in figure 4.62. It consists of the sputtering system, a tunable dye laser and a detection system. The cylindrical vacuum chamber had diameter of 300 mm, height of 360 mm, and was made of stainless steel. The chamber had ports for the target electrode, the pump, entrance and exit windows for the laser beam, and for an observation window for the fluorescence light. The target electrode, shown in figure 4.34, contained a permanent magnet that produced a ring-shaped magnetic field with maximum strength of about 0.05 T. The ratio of the electron cyclotron frequency $\omega_{ce} = 8.8 \times 10^9$ rad s^{-1} and the electron-neutral collision frequency $\nu_{en} = 2 \times 10^7$ Hz, was such that $\omega_{ce}/\nu_{en} \gg 1$, and so the condition for magnetron sputtering was well satisfied.

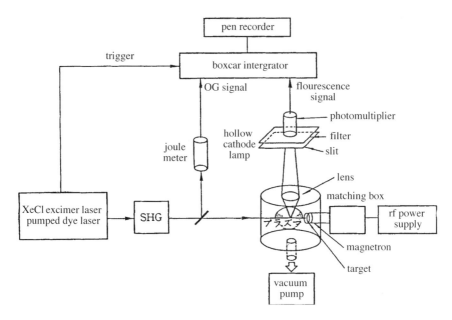

Figure 4.62. Arrangement for a LIF measurement of the velocity distribution function of sputtered particles in a magnetron discharge.

The target electrode was made of iron with purity of 99.9% and iron oxide. The sputtering was localized by placing a 5 mm circular aperture in front of the iron electrode. The optical arrangement for the experiment meant that the velocity component perpendicular to the electrode was measured.

The laser was a dye laser (Lambda-Physik, FL3002E) pumped by a XeCl excimer laser (Lambda-Physik, EMG-203LMSC) operated at 200 Hz. The laser was equipped with a prism, beam expander and grating, and an etalon was inserted into the laser oscillator so that the spectral width at the second-harmonic wavelength was 1.5 pm.

The fluorescence light was collected by a lens, with effective diameter of 56 mm and focal length of 100 mm, in the direction perpendicular to both the discharge axis and the laser beam. A slit (12 mm × 6 mm) and a narrowband interference filter placed in front of the photomultiplier detector reduce the signal from other sources of light. The filter had transmission of 35% at its central wavelength of 384 nm, and transmission spectral width of 24 nm. The photomultiplier was a HTV-R562 model from Hamamatsu Photonics.

Figure 4.63 shows part of the energy level diagram for atomic Fe. Excitation on the a^5D_4–y^5D_4 transition at 302.064 nm was used. Fluorescence was observed on the y^5D_4–a^5F_s transition at $\lambda = 382.043$ nm. The saturation characteristics and the wavelength range of the fluorescence signal were measured before the main experiment.

The main aim of the measurement was to observe the effect of collisions

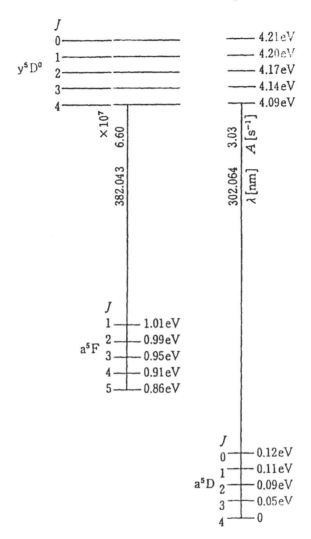

Figure 4.63. Partial energy level diagram for atomic iron.

between the atoms sputtered from the surface and argon atoms in the background gas. In order to observe this effect, two separate experiments were performed. In the first experiment, the position of the laser was fixed at a certain distance d from the electrode, and the fluorescence measured for a range of different pressures. In this experiment, the distance was 10 mm, and the pressure range was 1.3–33 mTorr. The measured spectra are shown in figure 4.64. In the second experiment, the pressure was held constant and the fluorescence measured as a function of distance from the target electrode. For this experiment, the pressure was 2.7 mTorr and the measured positions were $d = 10$,

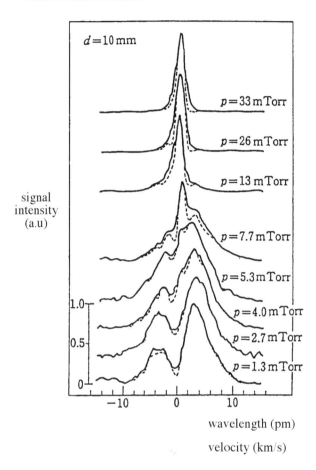

Figure 4.64. Velocity distribution functions measured at the same distance from the target electrode in plasmas with different gas pressures. The solid line shows the actual measured data while the dashed line shows the velocity distribution functions after deconvolution.

30 and 50 mm. These spectra are shown in figure 4.65. In both figures, measured spectra are shown as the solid lines. Each measured spectrum, however, is a convolution of the actual velocity distribution function and the spectral profile of the laser. The dotted lines in the figure show the velocity distribution functions determined by deconvolving the measured spectra.

The zero point for the velocity distribution function was determined by the signal from a hollow cathode lamp that had an iron electrode. In this experimental arrangement, a Doppler shift of 1 pm is equivalent to Fe atom velocity of 1 km s^{-1}. Fluorescence light also was detected from the laser beam after it had reflected from the surface of the target electrode. This appears in each figure as the 'negative' part of the spectrum. This provided an alternative

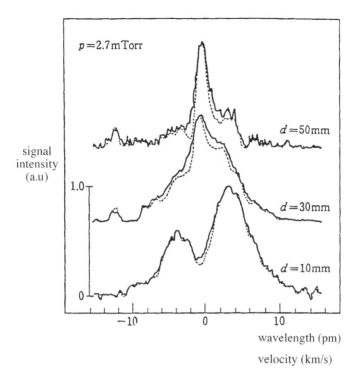

Figure 4.65. Velocity distribution functions measured at the different distances from the target electrode in plasmas with the same gas pressure. The solid line shows the actual measured data while the dashed lines show the velocity distribution functions after deconvolution.

means of determining the zero point of the velocity distribution function, and confirmed that the zero point from the optogalvanic signal was determined correctly.

At low pressure and for distances close to the electrode, the experimental results show that the velocity distribution function of the sputtered atoms was the well-known Thompson distribution [52]. At higher pressures, however, and for distances further from the electrode surface, a peak appears in the spectrum close to the zero point of the distribution. These results indicate that the change in the velocity distribution function depended on the product of pressure and distance pd, and further analysis was performed to examine this dependence.

The first step in this further analysis was to consider the spectra measured at low pressures and/or small distances from the electrode. The velocity distribution functions obtained from these spectra, which show few thermalization effects, were fitted with best-fit curves. An example of this curve-fitting procedure is shown in figure 4.66. The solid line is the experimental profile, obtained by deconvolving the original spectrum. The dashed curves are profiles calculated assuming a Thompson distribution, with each curve

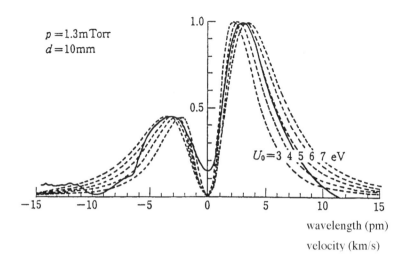

Figure 4.66. Comparison of a measured velocity distribution function with those fitted to the data assuming a Thompson distribution with different values of U_0. The measurement was made 10 mm from the target electrode in a plasma with gas pressure of 1.3 Torr.

representing the profile for a different value of U_0, the surface binding energy. The curve-fitting procedure indicated that the experimental profile was best fitted by a value of $U_0 = 5 \pm 1$ eV. This is in good agreement with the value of $U_0 = 4.3$ eV for pure iron.

The next step in the analysis procedure was to examine the way in which the measured profiles depended on pd, or the equivalent parameter of d/λ where λ is the mean free path of sputtered particles. These results are shown in figure 4.67. The vertical axis is $N_{\text{Maxwell}}/N_{\text{total}}$, which is the ratio of the number of thermalized atoms to the total number of atoms. This ratio indicates the proportion of the atoms that can be represented by a Maxwellian distribution. Each of the deconvolved profiles shown in figures 4.64 and 4.65 can be seen as a combination of a Thompson distribution and a Maxwellian distribution, and the ratio shown in the figure was calculated by determining the ratio of the area of the Maxwellian component of the distribution to the total area of the distribution. The value of λ used in the horizontal axis was calculated assuming gas temperature of 300 K. From the figure, it can be seen that the sputtered atoms have a mainly Thompson distribution for $d/\lambda \ll 1$ and that the Maxwellian component of the distribution becomes significant as $d/\lambda \to 1$.

The temperature of the thermalized Fe atoms could not be determined because the width of the Maxwellian distribution was much narrower than the spectral width of the laser.

These results have practical importance for sputtering experiments. For example, if the initial energy of the sputtered atoms and the distance between the target and the substrate is known, then the energy distribution function of sputtered atoms arriving at the substrate can be predicted. In addition, for a

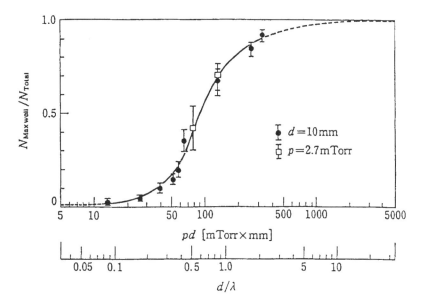

Figure 4.67. Ratio of the thermalized part of the velocity distribution function to the total distribution, as functions of p/d or d/λ.

given distance between the target and the substrate, the energy distribution function of the sputtered atoms arriving at the substrate can be controlled by varying the pressure of the background gas.

REFERENCES

[4.1] Griem H 1964 *Plasma Spectroscopy* (McGraw-Hill)
[4.2] Lochte-Holtgreven W 1968 *Plasma Diagnostics* (North-Holland)
[4.3] Adcock B and Plumtree W 1965 *Aero. Res. Council* C P 701
[4.4] Costley A, Chamberlain J, Muraoka K and Burgess D 1974 *Proc. Int. Conf. Submillimeter Waves (Atlanta)*
[4.5] van Milligen Ph, Soltwisch H and Lopes Cardozo N J 1991 *Nucl. Fusion* **31** 309
[4.6] Simonet F 1985 *Rev. Sci. Instrum.* **56** 664
[4.7] Costley E, Cripwell P, Prentice R and Sips A C C 1990 *Rev. Sci. Instrum.* **61** 2823
[4.8] Prentice R, Sips A C C, Fessey J A and Costley A E 1990 *Proc. EPS Conf. Control. Fusion Plasma Heating (Amsterdam)*
[4.9] Peacock N J, Robinson D C, Forrest M J, Wilcock P and Sannikov V V 1968 *Nature* **224** 488
[4.10] Johnson D, Dimock D, Grek B, Long D, McNeill D, Palladino R, Robinson J and Tolnas E 1985 *Rev. Sci. Instrum.* **56** 1015
[4.11] Salzmann H, Hirsch K, Nielsen P, Gowers C, Gadd A, Gadeberg M, Murmann H and Schrodter A 1987 *Nucl. Fusion* **27** 1925
[4.12] Gowers C, Fajemirokun H, Hender T, Kajiwara T, Nielsen P and Schunke B 1992 JET-IR(92)09

[4.13] Bindslev H, Hoekzema J A, Egedal J, Fessey J A, Hughes T P and Machuzak J S 1999 *Phys. Rev. Lett.* **83** 3206
[4.14] Takenaga H, Nakao T, Uchino K, Muraoka K, Maeda M, Iguchi H, Ida K, Yamada I, Okamura S, Yamada H, Morita S, Takahashi C and Matsuoka K 1995 *J. Nucl. Mater.* **35** 107
[4.15] Bay H L and Schweer B 1984 *J. Nucl. Mater.* **128** and **129** 257
[4.16] Honda C, Maeda M, Muraoka K and Akazaki M 1987 *Rev. Sci. Instrum.* **58** 759
[4.17] Kajiwara T, Shinkawa T, Uchino K, Masuda M, Muraoka K, Okada T, Maeda M, Sudo S and Obiki T 1991 *Rev. Sci. Instrum.* **62** 2345
[4.18] Matsuo K, Nitta H, Sonoda M, Muraoka K, Akazaki M and RFC-XX-M Group 1987 *J. Phys. Soc. Japan.* **56** 150
[4.19] Tanaka K, Matsuo K, Goto K, Bowden M, Muraoka K, Furukawa T, Sudo S and Obiki T 1992 *Japan J. Appl. Phys.* **31** 2260
[4.20] Kado S, Nakatake H, Muraoka K *et al* 1996 *J. Phys. Soc. Japan* **65** 3434
[4.21] JET 1993 *Annual Prog. Rep.*
[4.22] Doughty K and Lawler J E 1984 *Appl. Phys. Lett.* **45** 611
[4.23] Ganguly B N and Garscadden A 1985 *Appl. Phys. Lett.* **46** 540
[4.24] Preppernau B L and Ganguly B N 1993 *Rev. Sci. Instrum.* **64** 1414
[4.25] Greenberg K E and Hebner G A 1993 *Appl. Phys. Lett.* **63** 3282
[4.26] Moore A, Davis G P and Gottscho R A 1984 *Phys. Rev. Lett.* **52** 538
[4.27] Czarnetzki U, Luggenhölscher D and Döbele H F 1999 *Plasma Sources Sci. Technol.* **8** 230
[4.28] Doughty D K, Salih S and Lawler J E 1984 *Phys. Lett.* **103A** 41
[4.29] Maurmann S, Kunze H-J, Gavrilenko V and Oks E 1996 *J. Phys. B* **29** 25
[4.30] Derouard J and Sadeghi N 1986 *Opt. Commun.* **57** 239
[4.31] Yamagata Y, Suenaga K, Muraoka K, Maeda M and Akazaki M 1989 *Japan. J. Appl. Phys.* **28** 565
[4.32] Bowden M, Choi Y W, Muraoka K and Maeda M 1995 *Appl. Phys. Lett.* **66** 1059
[4.33] Kim J B, Kawamura K, Choi Y W, Bowden M and Muraoka K 1999 *IEEE Trans. Plasma Sci.* **27** 1510
[4.34] Lieberman M A 1988 *IEEE Trans. Plasma Sci.* **16** 638
[4.35] Kim J B, Kawamura K, Choi Y W, Bowden M, Muraoka K and Helbig V 1998 *IEEE Trans. Plasma Sci.* **26** 1556
[4.36] Gavrilenko V, Ikutake T, Kim H J, Kim J B, Bowden M and Muraoka K 2000 *Phys. Rev. E* to be published
[4.37] Mandich M L, Gaebe C E and Gottscho R A 1985 *J. Chem. Phys.* **83** 3349
[4.38] Yamagata Y, Kawano Y, Muraoka K, Maeda M and Akazaki M 1991 *Japan. J. Appl. Phys.* **30** 166
[4.39] Bowden M, Nakamura T, Muraoka K, Yamagata Y, James B W and Maeda M 1993 *J. Appl. Phys.* **73** 3664
[4.40] Akazaki M, Muraoka K and Hamamoto M J. 1981 *Inst. Elect. Eng. Japan.* **A101** 25 (in Japanese)
[4.41] Akazaki M, Muraoka K and Uchin K 1983 *J. Inst. Elect. Eng. Japan.* **A103** 609 (in Japanese)
[4.42] Itsumi Y, Uchino K, Muraoka K and Akazaki M 1984 *J. High Temp. Soc. Japan.* **10** 278 (in Japanese)
[4.43] Yamakoshi H, Kato M, Uchino K, Iwata T, Masuda M, Muraoka K, Maeda M and Akazaki M 1989 *Japan. J. Appl. Phys.* **28** L1589
[4.44] Uchino K, Kubo Y, Dozono H, Yamakoshi H, Muraoka K, Maeda M, Takahashi

A and Kato M 1995 *Appl. Phys.* **B61** 165
[4.45] Sakoda T, Momii S, Uchino K, Muraoka K, Bowden M, Maeda M, Manabe Y, Kitagawa M and Kimura T 1991 *Japan. J. Appl. Phys.* **30** 1425
[4.46] Bowden M, Okamoto T, Kimura F, Muta H, Uchino K, Muraoka K, Sakoda T, Maeda M, Manabe Y, Kitagawa M and Kimura T 1993 *J. Appl. Phys.* **73** 2732
[4.47] Cronrath W, Bowden M, Uchino K, Muraoka K, Muta H and Yoshida M 1997 *J. Appl. Phys.* **81** 2105
[4.48] Hori T, Bowden M, Uchino K and Muraoka K 1996 *Appl. Phys. Lett.* **69** 3683
[4.49] Hori T, Kogano M, Bowden M, Uchino K and Muraoka K 1998 *J. Appl. Phys.* **83** 1909
[4.50] Spitzer L Jr 1962 *Physics of Fully Ionized Gases* (New York: Interscience)
[4.51] Park W Z, Eguchi T, Honda C, Muraoka K, Yamagata Y, James B W, Maeda M and Akazaki M 1991 *Appl. Phys. Lett.* **58** 2564
[4.52] Thompson M W 1968 *Philos. Mag.* **18** 377

Chapter 5

Combustion Measurements

5.1 Combustion Fields and Laser-Aided Measurements

Understanding of phenomena that occur in combustion processes and internal combustion engines is important from the viewpoint of efficiency improvement, energy conservation and environmental protection. However, phenomena in *combustion fields* are extremely complicated. The mixture and flow of air and fuel, phase changes in the fuel, and the chemical reactions that lead both to formation of new species and generation of heat, are all inter-related phenomena [1]. In addition, these phenomena have sharp temporal and spatial dependencies. For all these reasons, it is difficult to understand the fundamental aspects of combustion processes.

Figure 5.1 shows laser-aided measurement techniques used in combustion research. In this figure, the main elements of combustion processes are divided into three main areas: the radicals and small particles that are produced by chemical reactions; the interaction between the fuel and the background air mixture, and the heat generated by the combustion process.

Measurement of a variety of quantities is desirable in order to understand the phenomena that occur during combustion. In particular, measurements of the following three quantities are important:

(i) concentration of each type of species (atoms, molecules, radicals, small particles);
(ii) temperature;
(iii) flow speeds of each species.

Furthermore, because of the complex environment, the measurements must satisfy the following requirements:

(i) high temporal resolution: because all of the above properties (concentration, temperature, and particle velocity) change rapidly;
(ii) high spatial resolution: two- and three-dimensional measurements are preferable;
(iii) non-perturbing;

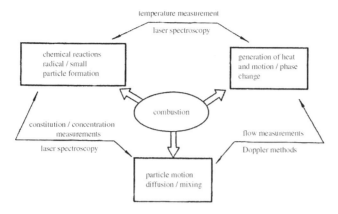

Figure 5.1. Combustion phenomena and laser-aided measurement methods.

(iv) *in-situ*: the measurements should be made in a high-temperature environment such as inside an engine or boiler.

Measurement techniques used until recently were not performed *in situ*, and only temporally averaged values of the combustion properties could be measured. Hence, the general understanding of combustion processes still is relatively poor. However, it should be clear from the discussion contained in previous chapters that laser measurement methods are capable of satisfying all of these demands. In this field, two- and three-dimensional, rather than single point, measurements are highly desirable. In this respect, imaging techniques that use lasers and CCD cameras are especially effective, and these types of measurements are being introduced rapidly into this research field.

5.1.1 Measurement of Particle Densities

The main techniques used for measurement of the density of the different particle species that exist in combustion environments are summarized in table 5.1. The first techniques listed are *Rayleigh scattering* and *Mie scattering*. These are examples of elastic scattering, and so the incident laser light and the scattered light have the same wavelength, and the scattered intensity does not depend strongly on the laser wavelength or the type of scattering particle. These properties mean that it is difficult to identify different types of particles with these techniques.

Rayleigh scattering is scattering from very small particles, with atomic and molecular size. This is the main type of scattering that occurs in combustion experiments performed under ordinary atmospheric conditions, with the scattering mainly arising from the O_2 and N_2 molecules in air. For this reason, the intensity of Rayleigh scattering usually depends linearly on the gas pressure.

Mie scattering is scattering from particles whose size is comparable with

190   *Combustion Measurements*

Table 5.1. Main particle measurement techniques for combustion study.

Measurement technique	Measured species
Rayleigh scattering	N_2, O_2 (pressure)
Mie scattering	small particles (particle radius 0.1–10 μm)
Raman scattering	N_2, O_2, CO_2
	combustion gases (H_2, CH_4, CO)
Laser-induced fluorescence (LIF)	OH, NO, H, O, CH, C_2, CN
Infrared absorption spectroscopy (IRAS)	same as LIF, exhaust gases
Nonlinear methods:	
coherent anti-Stokes Raman	
spectroscopy (CARS)	N_2, O_2, CO_2, CO
degenerate four-wave mixing (DFWM)	OH, NH_3, H_2, CO, H_2O, O_2, NO_2, NH

the incident light wavelength. In practical measurements, this means particles with radii in the range 0.1–10 μm. In the theory of scattering from small spheres, developed by Mie, the intensity of the scattered radiation is a complex function of the particle size, but, in general, the scattering intensity can be considered to increase with the particle size. Care has to be taken, though, when there is a distribution of particle sizes, because in this case, the relationship between particle size and the scattered light intensity is complicated. In addition, it can be very difficult to apply Mie scattering techniques for measurement of very small particles, often called clusters, that have nanometre order size, because the very weak scattered signal is difficult to detect. In higher vacuum conditions, mass analysis can be used but sometimes this method cannot be used in combustion environments.

The scattered signal intensity for both Rayleigh and Mie scattering increases as the incident light wavelength decreases, and high-power cw visible lasers such as Ar-ion lasers are commonly used as laser sources for these scattering measurements. Both of these types of scattering methods are used more often for temperature and velocity measurements, described in later sections, than for density measurements.

Raman scattering is a form of inelastic scattering from molecules. The frequency of the scattered light is shifted from that of the incident light by an amount that corresponds to a change in the rotational and vibrational energy of the molecule. The frequency shift is a characteristic property of the scattering particle, and so the scattering species can be identified by the magnitude of this frequency shift. Experimentally, the longer-wavelength (Stokes) components usually are measured. The cross-section for Raman scattering is small, and so in combustion environments, Raman scattering is normally used for measurements of species that have high concentrations, such as N_2, O_2, CO_2 and the combustion fuel itself. In these measurements, it is also often possible to measure the rotational temperature of the scattering species by detecting the different rotational components of the fundamental vibrational energy level. This procedure is explained in detail in section 5.2.2 .

There are a variety of laser spectroscopic techniques based on tunable lasers that can be used for measurement of atomic and molecular properties, but the most popular method for measurements in combustion environments is *laser-induced fluorescence* (LIF). Measurements of single atomic species, such as O and H, and diatomic species, such as OH, NO, CO and CH, can be made with high detection sensitivity. Many of the reported LIF measurements are of the OH radical, because OH is necessarily present in combustion. Even in ordinary atmospheric conditions, strong LIF signals can be obtained from OH radicals by using UV excitation. In addition, it is possible to determine rotational temperature by measuring the distribution of particle density among the rotational sub-levels of the ground vibrational state.

Another important feature of LIF is that one-dimensional and two-dimensional imaging measurements are feasible because of the high sensitivity and the large signals that can be obtained. In principle, all of the techniques discussed in chapter 7 for laser processing applications can be applied in exactly the same way in combustion research. Many of the atoms and molecules listed in tables 7.5 and 7.6 are important in combustion, and these can be detected by LIF techniques.

Infrared absorption spectroscopy (IRAS) is a useful technique that can detect various molecules that cannot be measured with LIF techniques. It rarely has been applied in combustion research, however, because of its relatively poor sensitivity and the lack of tunable laser sources for the infrared wavelength region. It is possible to use it for measurements of exhaust gases obtained by gas sampling because optical cells with long laser paths can be used for these measurements. For exhaust gas measurements, methods such as *Fourier transform infrared* (FT-IR) spectroscopy are practical and widely used. Infrared absorption techniques should become feasible for combustion research when practical, compact, tunable infrared laser sources become available.

One feature of laser-aided measurement methods in combustion environments is that *nonlinear spectroscopy* often is practical. The main nonlinear optical methods are *coherent anti-Stokes Raman spectroscopy* (CARS) and *degenerate four-wave mixing* (DFWM). Both these techniques have the common feature that, as a result of four-wave mixing produced by the third-order nonlinear susceptibility $\chi^{(3)}$, coherent light is radiated in a single particular direction. Very large amounts of background light usually are present in combustion environments, but by using CARS and measuring the coherent signal beam, signal-to-noise ratios that are much higher than for conventional Raman scattering can be obtained.

The CARS technique uses a four-wave mixing scheme such as that shown in figure 5.2 [2,3]. In the usual stimulated anti-Stokes Raman process, the ω_3 wave is generated by pumping the ω_1 wave. In CARS, a second laser is used to introduce a wave with frequency ω_2, corresponding to the Stokes frequency ($\omega_2 = \omega_1 - \omega_R$, where ω_R is the Raman frequency). A beam with frequency ω_3

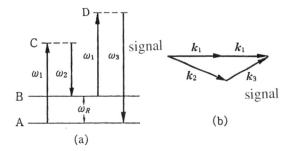

Figure 5.2. (a) Four-wave mixing scheme used for CARS measurements, and (b) the phase matching condition.

is generated as the signal wave. Different rotational levels of the molecules can be detected by scanning the frequency of the ω_2 beam. The signal beam has frequency of

$$\omega_3 = 2\omega_1 - \omega_2 \tag{5.1}$$

and the wavevectors of the three incident beams k_1, k_2, and k_3, must satisfy the following *phase matching* condition:

$$k_3 = 2k_1 - k_2. \tag{5.2}$$

This condition is shown in figure 5.2(b). The intensity of the signal beam I is given by

$$I \propto (\chi^{(3)})^2 I_1^2 I_2 \tag{5.3}$$

where I_1 and I_2 are the intensities of the ω_1 and ω_2 beams.

In CARS, the incident beams and the signal beam propagate in different directions, and so it is possible to detect only the signal beam by using a slit in front of the detector to block other incident light. For most cases, however, the difference in beam directions is extremely small. In this case, a collinear arrangement in which all three beams propagate in the same direction usually is used, and the signal beam is separated from the incidence beams by a wavelength filter. In this collinear arrangement, the coherence length corresponding to $\Delta k = 2k_1 - k_2 - k_3$ becomes important, and the signal intensity is only generated over a distance equivalent to this coherence length.

Although it is not necessary to use a tunable laser for the ω_1 wave source, the use of separate tunable lasers for both the incident beams allows a resonance effect to be used to increase $\chi^{(3)}$. This is done by making the two intermediate levels (levels C and D in figure 5.2(a)) resonant with electronic energy levels in the molecule, which makes $\chi^{(3)}$ and hence the signal beam intensity much larger. This technique is called *resonant CARS*.

The spectral resolution in a CARS measurement is determined ultimately by the spectral width of the ω_2 beam. However, by deliberately using a laser with a wide spectral width, and detecting the signal beam using a spectrometer

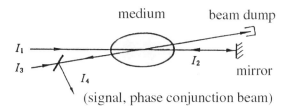

Figure 5.3. Generation of the phase conjugate signal beam I_4 in degenerate four-wave mixing.

and a diode array, it is possible to measure extremely fast single events such as explosions.

Degenerate four-wave mixing (DFWM) originally was devised as a method for obtaining *phase conjugation* of light [4]. A phase conjugate wave is a wave that has its spatial phase component reversed with respect to the original source wave. The phenomena of phase conjugation can be observed using many nonlinear optical effects. In DFWM, the third-order nonlinear susceptibility $\chi^{(3)}$ is increased greatly due to resonance effects. Figure 5.3 shows one possible arrangement for DFWM. The output beam I_1 from a tunable laser is passed through the sample gas to be measured, and then reflected from a mirror to produce a second beam I_2, that passes straight back through the sample along the same beam path. A third beam I_3, called the probe beam, is sent into the sample at a specific angle so that all three beams overlap simultaneously. The interaction of these beams with the nonlinear susceptibility causes a phase conjugate beam I_4 to be generated. This beam, the signal beam, propagates along the same optical path as the probe beam I_3 but in the opposite direction. All four beams, I_1, I_2, I_3, and I_4, have the same frequency. The intensity I_4 depends on the molecular density N, and is given by

$$I_4 \propto (\chi^{(3)})^2 I_1 I_2 I_3 N^2. \tag{5.4}$$

Compared with CARS techniques, DFWM methods require only a single tunable laser, and have greater sensitivity because the signal is generated by a resonance effect combined with a nonlinear effect. In both cases, the signal is generated as a coherent beam.

5.1.2 Measurement of Temperature

Temperature is the most important parameter to be measured in studies of combustion. For many years, the standard method of temperature measurement in combustion research involved comparing the measured emission spectrum with blackbody emission spectra. Active measurement of temperature using laser methods has led to large improvements in spatial and temporal resolution.

Several different techniques can be used for temperature measurement. The simplest method is Rayleigh scattering, and there are two separate ways of determining the gas temperature using this technique. One method is to measure the Doppler-broadened spectral profile of the scattered light. This broadening is cause by thermal motion of the gas particles, and the temperature can be determined from this broadening.

In most circumstances, gas particles with mass M and temperature T will have a spectrum with a Gaussian shape, given by the following expression:

$$g(\nu) = K \exp\left(\frac{-4 \ln 2 (\nu - \nu_0)^2}{(\Delta \nu_D)^2}\right) \quad (5.5)$$

where K is a constant, ν_0 is the central frequency and $\Delta \nu_D$ is the half-width determined by thermal motion. This width is given by

$$\Delta \nu_D = 2\nu_0 \sqrt{\ln 2 \frac{2kT}{Mc^2}}. \quad (5.6)$$

The Doppler-broadened width given by this expression is applicable not only for Rayleigh scattering spectra, but also for absorption and fluorescence spectra. Hence, in all the cases, the temperature T can be determined by measuring the spectral profile. However, the Doppler width is usually *extremely* small. Expressed in terms of wavelength, it is of picometre order. High-resolution spectral techniques are required to measure this width.

In the other Rayleigh scattering method, the scattered intensity I_R, which is proportional to the particle density N, is used to determined N, and then the ideal gas equation is used to determine the temperature [2]. The intensity I_R is given by

$$I_R = C I_0 N \sum_i (f_i \sigma_i) \quad (5.7)$$

where I_0 is the incident laser intensity, C is a constant and f_i and σ_i are the mole fraction and Rayleigh scattering cross-section for the different particle species in the gas. By assuming constant pressure and using the ideal gas equation, (5.7) can be written in terms of the gas temperature T to give

$$I_R = C' \sum_i (f_i \sigma_i) / T. \quad (5.8)$$

Although this is a seemingly simple and convenient method for determining T, the assumptions of constant pressure and constant $f_i \sigma_i$ often are not valid. Hence, although this technique is useful in some circumstances, there are many situations where it cannot be used.

In addition to Rayleigh scattering, various laser spectroscopic methods can be used to measure gas temperature. These methods include LIF, Raman spectroscopy, CARS, DFWM and laser absorption. In each of these methods, a portion of the vibrational/rotational spectrum of the gas molecules is measured, and the temperature is determined from the spectral profile. The total set of energy levels for a molecule, including rotational and vibrational levels in addition to the usual electronic levels, can be represented conceptually by the

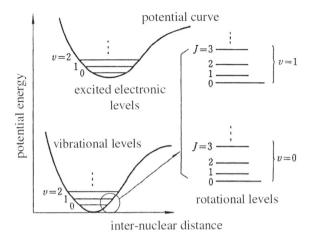

Figure 5.4. Schematic diagram showing vibrational and rotational energy levels in molecules

kind of diagram shown in figure 5.4. In this figure, each energy level is shown with a vibrational quantum number v_n and a rotational quantum number J_n. This figure shows the diagrams for a relatively small, simple molecule: for many-atom molecules, the diagram can become extremely complicated. Transitions between the levels are governed by a set of selection rules. The sets of vibrational/rotational transitions called the P, Q and R branch transitions are formed from these selection rules.

For the measurement of temperature, the population of molecules in the various vibrational and rotational sub-levels of the ground state is assumed to have a Boltzmann distribution. For vibrational levels, the density n_v of molecules in the vth vibrational level is given by

$$n_v = n\, g_v \exp(-E_v / kT) \tag{5.9}$$

where E_v is the energy and g_v is the degeneracy of the vth level. The density n_J of molecules in the Jth rotational level is given by a similar expression

$$n_J = n_v\, g_J \exp(-E_J / kT). \tag{5.10}$$

By combining these expressions, the relationship between temperature and the distribution of molecules in the vibrational/rotational energy levels can be expressed as

$$n_{vJ} = n g_v g_J \exp[-(E_v + E_J)/kT]. \tag{5.11}$$

For cases when it is possible to spectrally separate individual vibrational and rotational transitions, the temperature T can be determined by observing the intensify ratio of at least two lines in the vibrational/rotational spectrum. In most cases, however, many lines rather than just two are observed usually in order to obtain better accuracy. The temperature then is determined by plotting the distribution of intensities as a function of the energy of the lower state. For cases when it is not possible to observe individual lines, it is still possible to determine the temperature by comparing the measured spectral profile with a set of theoretically calculated profiles, each representing the profile expected for a given gas temperature.

Compared with the density measurements discussed previously, temperature measurements require relatively large signals because the signal-to-noise ratio must be sufficiently high over the whole spectral range being measured. For this reason, temperature measurements are usually made using gas species with high concentrations, such as N_2, O_2, H_2 and OH.

5.1.3 Measurement of Velocity

A laser diagnostic technique called *laser Doppler velocimetry* (LDV) is the most widely used technique for measuring the flow velocity of fluids, and this method also is used in combustion research. The laser method has many advantages compared with traditional methods such as thermal anemometers. Some of the advantages, such as its non-perturbing nature, its fast response and its high spatial resolution are common to most laser-based measurement methods. In addition, LDV has the specific advantages of being applicable to multi-phase flows and having a wide dynamic range for velocity measurements, being able to measure in the range 10^{-3}–10^2 m s^{-1}.

The theory and apparatus for LDV are described in detail in chapter 6. The most effective method is differential interferometry, in which two beams are crossed in the measurement region, forming interference fringes. Mie scattering from particles moving through this region is observed. In some cases, it is possible to use particles, such as oil drops and soot, that already exist in the combustion region as the scattering particles for the measurement. In most cases, however, the particles are introduced externally.

In addition to LDV, other methods discussed in chapter 6, such as holographic interferometry and pulse shadowgraphy, can be used to make two-dimensional measurements in the flow region. By using two-dimensional pulsed LIF methods, it is possible to observe the motion of particular species of atoms and molecules.

5.2 Examples of Combustion Measurements

A great number and wide variety of laser-aided measurements have been reported in combustion-related research. These range from experiments exploring fundamental aspects of combustion processes in burner flames to *in-situ* measurements in real combustion systems. In this section, representative examples of these measurements will be described, concentrating on density and temperature measurements made using LIF, CARS and DFWM techniques.

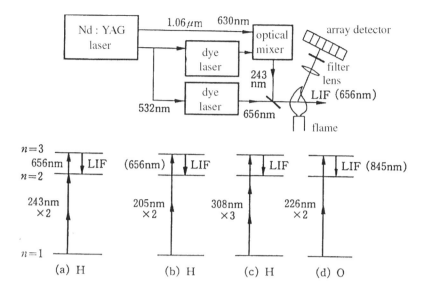

Figure 5.5. (*a*)–(*c*) Multiphoton excitation LIF methods for detection of hydrogen atoms [5], and (*d*) a scheme for oxygen [6].

5.2.1 Measurements by Laser-Induced Fluorescence Spectroscopy

One of the principal advantages of LIF techniques is their high detection sensitivity. When applied to combustion research, it is possible to detect even those particle species that exist in very small quantities. Tables 7.5 and 7.6 show a list of species that can be measured in laser processing research, and the list for combustion measurements is virtually the same.

An example of a relatively early LIF measurement is that of Goldsmith *et al* who measured H and O atoms in an H_2-air flame [5,6]. Excitation of light elements such as these is difficult with single-photon excitation because this requires radiation at vacuum-ultraviolet wavelengths. This radiation is absorbed by species in the atmosphere and so does not propagate through air. Hence, as was the case in this experiment, multi-photon excitation usually is performed.

In the Goldsmith experiment, detection of H atoms was performed using the scheme shown in figure 5.5(*a*). Initial excitation to $n = 2$ was achieved by two-photon absorption of 243 nm light, and further excitation from $n = 2$ to $n = 3$ was achieved by absorption of 656 nm light. Fluorescence detection was at the same Balmer-α wavelength of 656 nm. Although there is direct fluorescence from the $n = 2$ level, at the Lyman-α wavelength of 121.5 nm, this VUV wavelength does not propagate through the atmosphere, and so this direct fluorescence light could not be detected.

The schemes shown in figures 5.5(*b*) and (*c*) each require a single tunable laser to achieve excitation to $n = 3$. It has become possible recently to generate short wavelength radiation by using BBO crystals to generate the second-

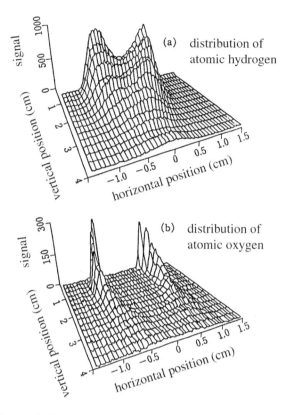

Figure 5.6. Measured distributions of hydrogen and oxygen atoms in flame, obtained using multiphoton excitation LIF techniques.

harmonic wave of the fundamental output of a laser, but it is still difficult to efficiently generate wavelengths in the region of 205 nm. In order to generate this wavelength, the most common method is to use an ArF laser ($\lambda = 193$ nm) combined with a Raman wavelength shifter [7]. The three schemes discussed so far are appropriate for detection of H atoms. In the Goldsmith experiment, O atoms were also detected, using the two-photon excitation scheme shown in figure 5.5(d) [6].

The detector in this experiment was an image-intensified diode array. This enabled the measurement of a one-dimensional spatial distribution of the particle density. An example of a measured distribution is shown in figure 5.6. In this experiment, it was feasible to measure this spatial distribution using the fluorescence light from only one laser shot, but the signals shown in the figure were averaged over 50 laser shots in order to improve the signal-to-noise ratio. From the viewpoint of flame research, these measurement results are interesting because they indicate that the vertical distributions of H and O are quite different.

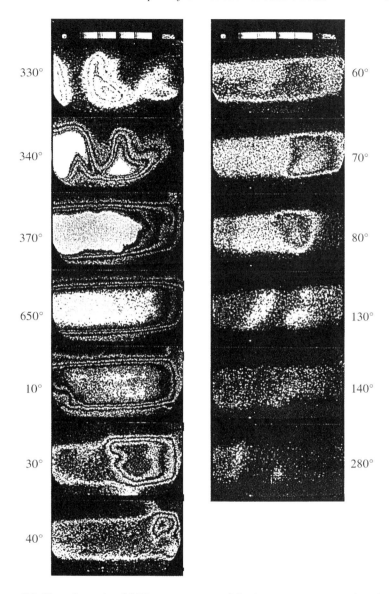

Figure 5.7. Two-dimensional LIF measurements of the *iso*-octane concentrations inside the cylinder of an internal combustion engine [8]. The angle shown in the crank angle, and ignition occurs at 714°.

LIF techniques also can be used to obtain two-dimensional distributions of particle densities. Figure 5.7 shows a result obtained from an LIF measurement of *iso*-octane (*iso*-C_8H_{15}) inside the cylinder of an internal combustion engine [8]. The angle shown in the figure is the crankshaft angle: 720° represents one full revolution and the rotation speed was 1000 rpm. The excitation was

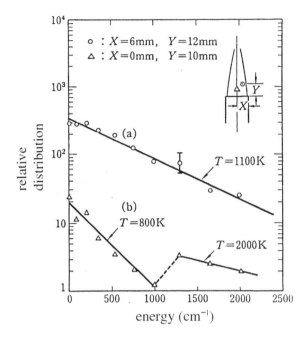

Figure 5.8. Examples of rotational temperature measurements obtained by LIF detection of OH [9].

achieved by a sheet laser beam formed from the output of a 248 nm KrF laser. An image-intensified CCD camera was used to detect the signal. The image intensifier was gated with a 200 ns gate pulse that was synchronized with the laser pulse. In the same experiment, OH and NO molecular distributions inside the cylinder were measured using the same apparatus.

The OH radical is the most frequently used species for measurement of temperature using LIF techniques. Figure 5.8 shows the result of temperature measurements in a flame, obtained using the R_1 branch transitions from the ground level of OH [9]. The measured transitions were the $^2\Pi(v=0) \rightarrow\ ^2\Sigma^+(v=1)$ transitions at $\lambda \sim 281$ nm. The intensity of the fluorescence from the kth level is given by

$$I_k = n_k \omega(R_1,k) \eta_F \qquad (5.12)$$

where η_F is the fluorescence quantum yield and $\omega(R_1,k)$ is the transition probability for the R_1 branch transition from the kth level. In order to calculate the temperature distribution, $\omega(R_1,k)$ and η_F must be known. In this measurement, η_F was assumed to be the same for each energy level, but this assumption is not valid for high-pressure situations. For high-pressure conditions, relaxation due to collisions and the effect of self-absorption can make the relationship between the LIF signal intensity and the molecular density extremely complicated [10]. The distribution shown in figure 5.8(a) is an

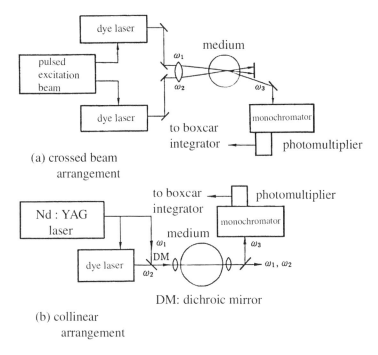

Figure 5.9. Experimental arrangements for measurements by coherent anti-Stokes Raman spectroscopy (CARS).

exponential function of the energy, and this means that it is valid to use equation (5.11) to determine the temperature T for this case. For the case of the distribution shown in figure 5.8 (*b*), however, it seems that there is a non-equilibrium distribution of molecules in the higher energy levels, and so the temperature cannot be determined directly in this simple way [9].

One of the major advantages of LIF is that it is often possible, due to the large signals, to use a CCD camera to make two-dimensional imaging measurements. It is even possible to determine a two-dimensional temperature distribution if the distribution of signals for a number of rotational levels can be imaged. For this, it is necessary to scan the excitation wavelength in order to measure the spectral profile, but such measurements of OH rotational temperature have been demonstrated using a rapid frequency-scanning laser [11,12].

5.2.2 Measurements by Coherent Anti-Stokes Raman Spectroscopy

Two typical experimental arrangements for coherent anti-Stokes Raman spectroscopy (CARS) are shown in figure 5.9 [3]. Part (*a*) of the figure shows the so-called *crossed-beam arrangement*, in which two dye lasers are used to generate the ω_1 and ω_2 frequency beams, and the beams are made to overlap inside the gas. Part (*b*) of the figure shows the *collinear arrangement*, in which

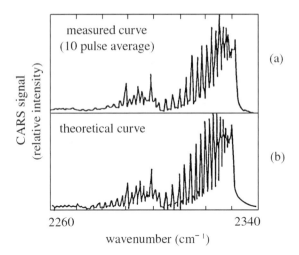

Figure 5.10. Comparison of measured and theoretical spectra for the Q-branch of N_2, measured by CARS in pre-mixed methane–air flame [13].

the ω_1 beam is provided by a fixed frequency Nd:YAG laser, and the two beams, ω_1 and ω_2, pass through the gas collinearly. In this arrangement, a filter is used to separate the ω_3 signal beam. The relatively simple collinear arrangement is well suited for measurements in gases that have low dispersion.

Nitrogen is the most common target molecule for CARS measurements of gas temperature. Figure 5.10 shows an example of a Q-branch N_2 spectrum measured in a CH_4–air flame using CARS. A theoretical spectrum, calculated by computer, also is shown [13]. This kind of measurement, in which the entire spectral profile is measured and then compared to a calculated profile, is the most reliable method of measuring temperature. It is important, however, that the calculation of the spectral profile is performed carefully because effects such as collisional narrowing of the spectrum at high pressures have to be considered [14]. The spectral width of the dye laser used in the experiment also affects the measured profile.

Figure 5.11 shows the experimental arrangement and results of a CARS measurement, made in the collinear arrangement, used to determined the temperature distribution inside a diesel engine [15]. The data shown are the average of repeated measurements when the crank angle was 368°. The spatial distribution was measured by translating the measurement point along the laser axis by changing the position at which the laser was focused. For this measurement, the collinear arrangement had to be used but better spatial resolution is usually possible with the crossed-beam arrangement, in which the spatial distribution can be measured by changing the crossing point of the beams [16].

In order to measure a full spectrum by CARS, it is necessary to scan the laser wavelength, and this results in relatively long measurement times. There are examples of single-shot CARS measurements [17], but the experimental

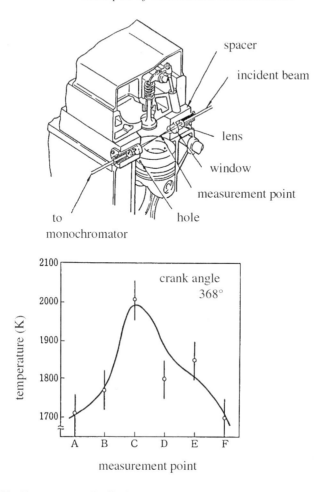

Figure 5.11. Temperature distribution inside a diesel engine, obtained by CARS measurements of N_2 using the collinear optical arrangement [15].

apparatus is different in several significant ways to that shown in figure 5.9. The experiment described in [17] used an experimental arrangement that was schematically similar to that shown in figure 5.9(a), but a broadband dye laser (width ~ 80 cm^{-1}) was used in place of the spectrally narrow dye laser, and a detector array with an image intensifier was attached to the spectrometer to measure the spectral profile. An example of the results obtained in this single-shot CARS measurement is shown in figure 5.12. The graph shows the measured N_2 Q-branch spectrum, together with a calculated profile that was fitted to the measured spectral profile.

CARS measurements require relatively complicated experimental apparatus, and this is a disadvantage compared with some other laser techniques. However,

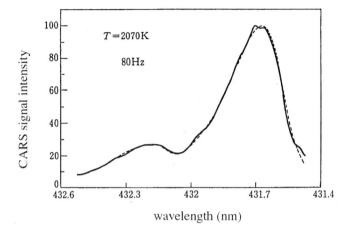

Figure 5.12. Example of a temperature measurement in a turbulent flame, obtained from a single-shot CARS measurement of N_2 [16]. The solid line is the measured data.

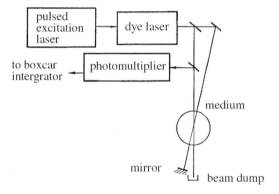

Figure 5.13. Example of a measurement system for degenerate four-wave mixing (DFWM).

the CARS method of determining gas temperature is the most reliable measurement method that currently exists. Although the uncertainty obviously varies from experiment to experiment, a typical value of the uncertainty for measurements in a flame with temperature of about 2000 K might be ±50 K.

5.2.3 Measurements by Degenerate Four-Wave Mixing

Degenerate four-wave mixing (DFWM) is a similar technique in many ways to CARS, but there are two advantages of using DFWM. The first of these is that the sensitivity is higher, because the signal generation mechanism is a combination of a nonlinear effect and a resonance effect. The second advantage is that the measurement system is simpler because only a single tunable laser is necessary.

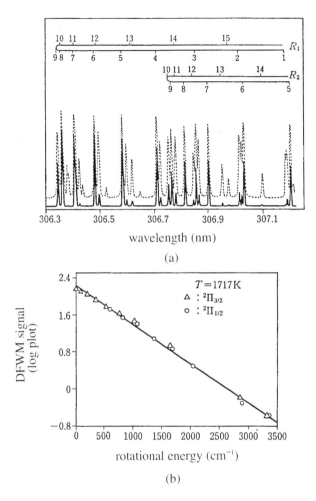

Figure 5.14. (a) R_1 and R_2 spectra of OH, obtained by DFWM (solid line) and LIF (dashed line), and (b) Boltzmann plot of the DFWM signals [17].

Figure 5.13 shows an experimental arrangement for a DFWM experiment. In this arrangement, the output from the tunable laser is divided into two beams that are crossed over in the medium at an angle of 4°. One beam is reflected by a mirror so that it propagates straight back along its original path. The other beam, called the probe beam, is directed into a beam dump. The DFWM signal is the phase conjugate of the probe beam, and so travels along the same path as the probe beam, but in the opposite direction. This beam is detected by using a half-mirror to split off the signal beam and direct it into a detector [17,18].

Because DFWM has high sensitivity, it can be used for detection of many species that are present in combustion. In order to explain the method in more detail, one example will be described: the measurement of temperature by the detection of spectra from OH radicals. Figure 5.14(a) shows the DFWM signal due to OH radicals in a propane–air flame, obtained by scanning the laser wavelength in the region of λ = 306 nm. The R_1 and R_2 branches of the $A^2\Sigma - X^2\Pi(0,0)$ transition were observed. In this case, the DFWM signal intensity for each transition is

$$I_J \sim [B_{ij}N(v,J)]^2 \tag{5.13}$$

where B_{ij} is the transition probability for the jth transition and $N(v, J)$ is the density of OH radicals in the (v, J) vibrational/rotational energy state. The distribution of particles within each vibrational state is given by

$$N(J) \sim (2J+1)\exp(-\Delta E_n hv / kT) \tag{5.14}$$

where ΔE_n is the energy difference of the $J = n \pm 1/2$ states. As can be understood by looking at these equations, the temperature can be determined by plotting the data on a graph with the rotational energy on the horizontal axis, and $J_n^{1/2}/B_{ij}$ [$2(n \pm 1/2)+1$] plotted on a log scale on the vertical axis. Figure 5.14(b) shows the data for R_1 and R_2 branches obtained from part (a) of the figure. The data can be well fitted by a straight line, and the temperature determined from the gradient of this line was 1717 ± 50 K.

Since the DFWM signal is coherent and intense, two-dimensional measurements are feasible. This can be achieved by expanding the beams so that a large cross-sectional area can be observed and then using a CCD camera to detect the signal. Such experiments currently are being attempted [19,20].

REFERENCES

[5.1]   Kohda S and Takubo Y ed 1990 *Flame Spectroscopy* (Gakkai Shuppan Centre) (in Japanese)
[5.2]   Levenson M D 1977 *Physics Today* **30** 44
[5.3]   Maeda S *et al* 1979 *Bunko Kenkyu* **28** 353 (in Japanese)
[5.4]   Dreier T and Rakestraw D J 1990 *Appl. Phys.* **50** 479
[5.5]   Goldsmith J E M and Anderson R J M 1985 *Appl. Opt.* **24** 607
[5.6]   Alden M *et al* 1984 *Appl. Opt.* **23** 3255
[5.7]   Kajiwara T *et al* 1991 *Rev. Sci. Insrum.* **62** 2345
[5.8]   Andersen P *et al* 1990 *Appl. Opt.* **29** 2392
[5.9]   Wang C C and Davis L I Jr 1974 *Appl. Phys. Lett.* **25** 34
[5.10 Cattolica R 1981 *Appl. Opt.* **20** 1156
[5.11] Chang A Y *et al* 1990 *Opt. Lett.* **15** 706
[5.12] Honda C *et al* 1991 *Japan. J. Appl. Phys.* **30** 72
[5.13] Ohtake K 1989 *Koongakkaishi* **15** 7 (in Japanese)
[5.14] Hartley D L 1988 *J. Quant. Spectrosc. Radiat. Transfer* **40** 291
[5.15] Kajiyama K *et al* 1982 *SAE Technical Paper Series* No 821036
[5.16] Lange B, Noda M and Marowsky G 1989 *Appl. Phys.* **49** 33
[5.17] Dreier T and Rakestraw D J 1990 *Opt. Lett.* **15** 72
[5.18] Mann B A *et al* 1992 *Appl. Phys.* **B54** 271
[5.19] Hemmerling B and Stampanoni-Panariello A 1993 *Appl. Phys.* **B57** 281
[5.20] Morishita K, Higuchi Y and Okada T 1998 *Japan. J. Appl. Phys.* **37** 4383

Chapter 6

Measurements in Gas Flow Systems

Gaseous flows are important parts of many industrial processes. They are important in machines such as fans, pumps and ventilators, which are major components of many factory systems, and in aviation, which has become remarkably widespread in recent years. A large amount of research is aimed at improving the efficiency of these machines, and, to a great extent, this research involves understanding basic hydro- or fluid-dynamic mechanisms that occur in these devices. Experimental methods for measuring quantities associated with the flow include measurement of the pressure at the walls around the flow and measurement of flow properties such as temperature using instruments inserted into the flow. These instruments, however, often perturb the flow or have poor temporal and spatial resolution. When such measurements can be made, it is sometimes difficult to analyse the measured data to obtain accurate information about the flow. For these reasons, accurate non-perturbing measurement techniques are desired. Optical diagnostic techniques are well suited for these measurements, but practical methods became possible only with the development of lasers and their associated optical technology.

The most common optical methods used for flow measurements measure the change in refractive index produced by changes in density of the flow. This class of density measurement methods is described in section 6.1, and includes methods such as schlieren techniques and interferometry, used in studies of shock waves in supersonic flows. Early measurements were performed with other light sources, but the use of lasers has led to great improvements in measurement qualities such as fringe intensity and contrast, and has greatly increased the effectiveness of these techniques. Flow velocity measurements are described in section 6.2. These methods are based on observation of Mie scattering from tracer particles in the flow. A new research area recently has been established based on laser-induced fluorescence techniques used for measurements of flow parameters and for flow visualization. This new area is discussed in section 6.3.

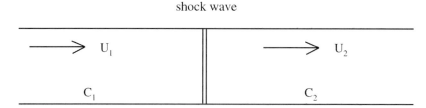

Figure 6.1. Schematic diagram showing gas flow across a stationary normal shock wave. The velocity u_1 upstream of the shock wave is supersonic and the velocity after the shock wave u_2 is subsonic. c_1 and c_2 are the speeds of sound in that part of the flow.

6.1 Measurement of Refractive Index Changes (Density Measurements)

In gas flow research, the most successful application of the density measurement techniques that detect changes in refractive index is the visualization of shock waves. These methods were used in the development of supersonic aircraft in the 1950s. Later, when high-speed compressors were developed in industrial machinery, these methods were used to understand the high-speed flows in these machines.

When gas travelling at supersonic speeds collides with a stationary object, or when a supersonic flow is induced in a stationary gas, the gas is decelerated to subsonic speeds in an extremely thin layer, called a *shock wave*. Many of the examples used in this chapter describe observations of shock wave phenomena and so it is appropriate first to consider the physical mechanisms that produce this extremely thin shock wave layer.

Figure 6.1 shows the case of a stationary normal shock wave. A supersonic flow, with flow speed u_1, is decelerated to a subsonic flow, with flow speed u_2, as it flows across a stationary shock wave that is vertical to the flow. The shock wave heats the gas, and so, in the downstream region, the speed of sound c_2 is higher than the flow speed u_2. This means that disturbances downstream from the shock wave can reach the wave. In the upstream region, however, the speed of sound c_1 is less than the flow speed u_1, and so upstream disturbances cannot reach the shock wave. This results in gas temperature and density gradients becoming progressively larger at the shock front. These sharp gradients can be balanced only by particle diffusion and heat transmission, which are driven by the density and temperature gradients. The balance between these processes determines the equilibrium temperature and density gradients in this region.

Both particle diffusion and heat transmission occur because of collisions in the gas, and so the redistribution of temperature and density occurs over a distance of the order of the mean free path of the gas particles. The distance determines the thickness of the shock wave. At atmospheric pressure and temperature, mean free paths are less than 1 μm, and so the shock wave consists of an extremely thin layer of gas. Inside this thin layer, temperature and density have extremely large gradients. The refractive index of gases is determined

Measurement of Refractive Index Changes

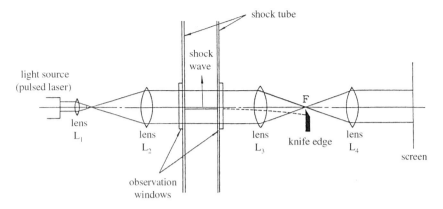

Figure 6.2. Schematic diagram of a schlieren apparatus for imaging of a shock wave.

solely by the density, and so the presence of the shock wave and its associated density gradients mean that there are correspondingly large gradients in the refractive index of the gas. Special optical methods have been developed to observe this phenomenon.

6.1.1 Schlieren Method

An example of a schlieren measurement system that was used to measure a shock wave travelling along a tube is shown in figure 6.2. A pulsed laser is used to probe the shock wave at the time it passes the observation windows. During the time of the laser pulse, the shock wave will move a distance of $L = \Delta \tau M a$, where $\Delta \tau$ is the pulse width of the laser, M is the Mach number of the shock wave and a is the speed of sound. (The Mach number M is the ratio of the shock wave velocity to the speed of sound.) In the example discussed below, the laser source was a Q-switched ruby laser, with pulse width $\Delta \tau = 20$ ns. In this case, an $M = 3$ shock wave would move a distance of $L = 20$ μm during the time of the laser pulse. This is larger than the width of the shock wave itself, but is not large enough to be resolved by the optical measurement described below. Therefore, in this case, the motion of the shock wave during a laser pulse could be neglected.

The principle of the schlieren method can be understood by considering what happens when a shock wave travels along the tube in the system shown in figure 6.2. In this apparatus, the laser beam is enlarged by cylindrical lens L_1 and then collimated by lens L_2 to form a parallel and sheet-shaped beam that passes through the tube. The beam is then transmitted to the screen via lenses L_3 and L_4. When a shock wave travels along the tube, part of the beam will pass through the shock wave and part of the beam will be unaffected. The part of the beam that does not pass through the shock wave will not be refracted, and hence will be

210 *Measurements in Gas Flow Systems*

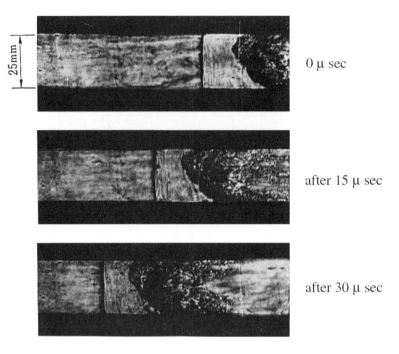

Figure 6.3. A schlieren photograph showing a flame front approaching a shock wave, to eventually form a detonation wave. Other features apart from the shock wave and the flame can be seen, but these are due to the refractive index of the observation window, which was made of Pyrex. (Photograph courtesy of Professor S Ono, Kyushu University.)

imaged at the focal point F of lens L_3, before being transmitted to the screen. The part of the beam that does pass through the shock wave, however, will be refracted, its propagation direction will change, and it will be focused at a point vertically below F, as shown in the figure. In the schlieren method, this refracted light is blocked by placing a knife-edge with its leading edge just below the focal point F. Hence, in the image viewed on the screen, regions corresponding to the position of the shock wave will be reduced in light intensity: the position of the planar shock wave is indicated by a dark line in the image.

Examples of photographs taken using such a system are shown in figure 6.3. The light source for this experiment was a stroboscopic ruby laser with flashlamp excitation, having a pulse width of 20 ns, and switched at 5 μs intervals (i.e. 200 kHz). A streak camera having a rotating mirror was used for the photography. In this experiment, a mixture of CH_4 and O_2 gases at 116.7 mTorr was ignited at a position on the right-hand side of the photograph. The dark line indicates the shock wave, and propagation of the flame behind the shock wave also can be seen. From these photographs, the speed of the shock wave was determined to be 1686 m s^{-1}, while that of the flame was determined to be 1141 m s^{-1}. The shock wave was generated by the compressive motion of the flame, and the flame approached the shock wave to eventually merge and form a so-called *detonation wave*.

In this example, the knife-edge was placed parallel to the refractive index gradient (i.e. to the density gradient). As shown by equation (2.26), the brightness of the image is proportional to the density gradient, and so, in principle, the spatial distribution of density can be determined from the spatial integration of the distribution of the brightness. In practice, however, the accuracy of this method usually is relatively poor. Hence, while the schlieren method is useful for obtaining information about the position and shape of shock waves, interferometric methods described below are used when quantitative information of density across a shock wave is required.

As discussed in section 2.2.2, the refractive index of gases is almost independent of wavelength (figure 2.12). However, there is a small wavelength dependence, and the density gradient across a shock wave is extremely large. These effects mean that if a white light source or a multi-line laser is used as the light source for a schlieren experiment, the amount of diffraction at the shock wave will be different at each wavelength. In this case, the knife-edge shown in figure 6.1 can be replaced by a three-colour filter and colour film can be used at the screen to record the image. An image obtained in this way shows the refractive effect of the shock wave front much more clearly than one produced using a monochromatic light source. This is called *the colour-schlieren method*.

6.1.2 Shadowgraphy

Shadowgraphy resembles the schlieren method in some ways, but is a simpler method for imaging shock waves. The simplest experimental arrangement is one in which the lens L3 in the schlieren apparatus shown in figure 6.2 is replaced by a screen. Both the light refracted by the shock wave and the light not affected by it are recorded simultaneously on the screen, producing a shaded distribution of light.

Figure 6.4 shows a photographic image of a shock wave image obtained using shadowgraphy. The light source was a Q-switched ruby laser with pulse energy of 0.1 J and pulse width of 20 ns. A device called a shock tube was used to generate the shock wave. This device is a pipe with a square or cylindrical cross-section that contains a membrane separating two regions of different pressure. The membrane is made of cellophane or thin aluminium, and there is a large pressure difference between the regions on either side of the membrane. If the membrane is suddenly destroyed, using methods such as a needle for cellophane or a wire discharge for aluminum, the sudden flow from the high-pressure region to the low-pressure region generates a shock wave. The Mach number of the shock wave depends on the kind of gas and the pressure ratio, but shock waves with Mach numbers of 2 or 3 can be produced easily with such a device.

In the experiment discussed here, CO_2 gas was used, with a pressure ratio of about 25. The pressure on the low-pressure side was 26 kPa (19.5 Torr) and

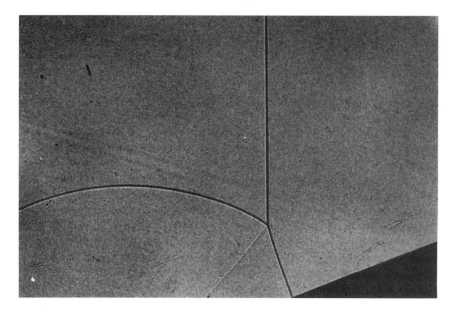

Figure 6.4. Image of a shock wave obtained using shadowgraphy [2]. The shock wave originated to the left of the figure, and propagated from left to right. The flow in the region in the lower side has a complicated structure. (Photograph courtesy of Professor K Matsuo, Kyushu University.)

the pressure on the high-pressure side was 650 kPa (6.4 atm), leading to the generation of a shock wave with Mach number estimated to be 2.0. A condition called the Rankine–Hugonoit relation states that, for a shock wave in thermal equilibrium, the ratio of thermodynamic quantities such as temperature and density in front of and behind the wavefront is determined solely by the type of gas and the Mach number. The density ratio across the normal shock wave shown in figure 6.4 is predicted by this relation to be 3.12.

In shadowgraphy, the light intensity in some parts of the image is reduced due to light being redirected, via refraction by the shock wave, to other parts of the image. Hence, some parts of the image become darker while other parts become brighter. The light intensity on a shadowgraph image is proportional to the second derivative of density, $\nabla^2 \rho$, as given in equation (2.27).

6.1.3 Interferometry

In plasma interferometry, both Mach–Zehnder type interferometers (figure 2.10), discussed in sections 4.2.1 and 4.3.2, and the Michelson interferometers (figure 2.11) are used widely. However, for measurements of shock waves, mainly the Mach–Zehnder arrangement is used. The reason for this is that the beam passes only once through the shock wave in the Mach–Zehnder arrangement, whereas a double pass is used in the Michelson arrangement. Because the beam is strongly

Figure 6.5. Photographic image of a shock wave obtained using an interferometric method [3]. The laser apparatus used for the measurement was the same as that for the image shown in figure 6.4. (Photograph courtesy of Professor K Matsuo, Kyushu University.)

refracted by the shock wave, a single pass is sufficient for the measurement. In addition, the Mach–Zehnder arrangement is easier to use in terms of setting up the system and recording the signal.

Figure 6.5 shows a photograph of a shock wave, similar to that used for the measurement shown in figure 6.4, obtained using an interferometric method. The change in density $(n_n - n_0)$ can be determined by counting the change in the number of fringes ΔN before and after the shock wave, and then using equation (4.5) with $n_i = n_e = 0$. For the case of the flow shown in figure 6.5, ΔN was determined to be 21, resulting in a value of n_n/n_0 of 3.11, calculated using a value of n_0, the density before the shock wave, of 6×10^{24} m^{-3}. The Rankine–Hugoniot relation for a shock wave with $M = 2.0$ predicts a value of $(n_n/n_0)_{RH}$ of 3.12, which agrees well with the measured value.

6.1.4 Holography

The imaging methods discussed above are useful in many circumstances, but are inadequate when the flow has a three-dimensional structure. In these cases, the method of holography can be used to obtain three-dimensional information

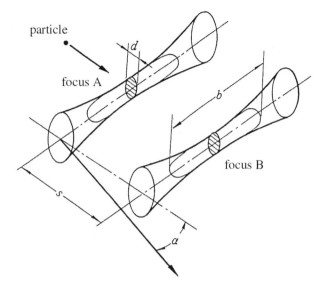

Figure 6.6. Direction of particle flow with respect to the laser beam foci A and B in an L2F measurement. (Courtesy of Professor H Hayami, Kyushu University.)

about the flow. One example of such a measurement is the observation of three-dimensional distributions of liquid fuel particles in an engine, performed using in-line holography.

At present, holographic imaging methods are not used widely but they can be expected to become more widespread as the need for information about complex flows in different machines increases.

6.2 Measurement of Flow Velocity

The basic principle behind laser-aided measurements of flow velocity is that a spatial distribution of laser intensity is generated in the flow region, and Mie scattering from dust in that region is observed. There are two methods for establishing this spatial distribution, called the laser 2 focus (L2F) method, and the interference (IF) method.

In the L2F method, two beams of light are focused so that the focal points of the beams are separated by a distance s, as shown in figure 6.6. In principle, it is not necessary to use a laser as the light source. However, an intense light source with good directivity is necessary to measure the flow velocity in the extremely narrow spaces inside a machine such as a pump or turbine. Hence, this measurement method became practical only after the development of lasers.

The principle of the IF method is shown in figure 6.7. A laser beam is divided into two parts, which are crossed over at an angle θ to generate interference fringes of high and low laser intensity in the intersection region.

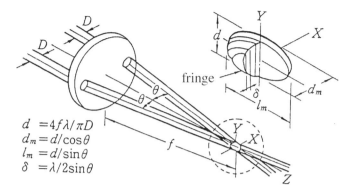

$d = 4f\lambda/\pi D$
$d_m = d/\cos\theta$
$l_m = d/\sin\theta$
$\delta = \lambda/2\sin\theta$

Figure 6.7. Arrangement of the beams in the observation region for a measurement using the IF method. This apparatus usually is called a laser Doppler velocimeter (LDV). (Courtesy of Professor H Hayami, Kyushu University.)

This method also became practical only with the development of intense, coherent laser sources. An instrument using this method to measure flow velocity usually is called a *laser Doppler velocimeter* (LDV).

The usefulness of laser-based methods of flow velocity measurement was immediately recognized after the development of lasers. The first instrument was reported in 1964 and there have been continual improvements since then. Presently, many models are commercially available and these are in widespread use.

6.2.1 Measurement Techniques

As mentioned above, the basic principle of these measuring techniques is measurement of light scattered from dust particles in the flowing gas. In cases where the amount of dust naturally present in the flow is insufficient, small dust-like particles are introduced into the flow specifically for the measurement. If the dust particles are too large, their flow velocity may differ from that of the gas, but if they are too small, the amount of Mie scattering might be too small for detection. Usually, dust particles with radii of the order of visible or a near-IR wavelengths, of the order of 0.1–10 µm, are used.

The following two sections contain general descriptions of the two methods.

L2F. The arrangement of the laser beams in an L2F experiment is shown in figure 6.6. There is no spatial resolution over the length b, which is determined by the focal depth of each beam. The dimensions of the system are usually of the order of $s = 0.2$–0.5 mm, $d = 0.01$ mm, and $b = 0.8$–1.0 mm. The beam is well focused which means that the scattering from L2F is about 100 times larger than that of LDV (discussed below) for dust particles of the same size. Another way of expressing this is to say that even small dust particles, of

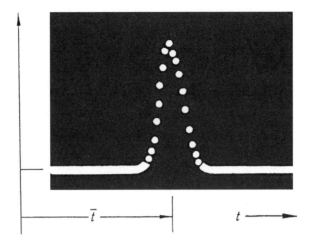

Figure 6.8. Distribution of signals with respect to time t in an L2F experiment (signal occurrence function). (Courtesy of Professor H Hayami, Kyushu University.)

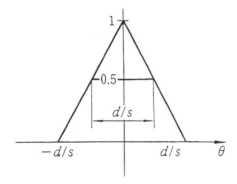

Figure 6.9. Schematic diagram showing the distribution of signals with respect to t in an L2F experiment (signal occurrence function). The distribution was calculated considering only d and s (i.e. neglecting turbulence). (Courtesy of Professor H Hayami, Kyushu University.)

0.1–0.2 μm size, can generate a sufficiently large scattering signal in an L2F measurement. For this reason, flows such as shock waves that have extremely sudden changes in velocity can be measured by using small dust particles that can respond quickly to flow changes.

In the L2F method, it is necessary to differentiate between signals from dust at B that has passed through A and signals from other dust. This is done by generating a signal occurrence function, such as that shown in figure 6.8, for particles passing through A, and eliminating results that differ significantly from this function.

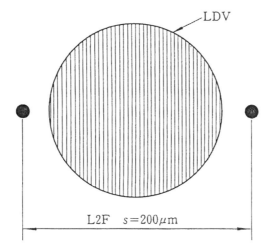

Figure 6.10. Light intensity distributions in the measurement regions of typical L2F and LDV experiments, showing the different dimensions of the two methods. (Courtesy of Professor H Hayami, Kyushu University.)

Only particles that flow along the line connecting the focal points A and B will contribute to the signal. The degree to which this determines the flow direction depends on the size d of the focal points at A and B, and the distance between the points, s. The signal occurrence function determined by these two parameters has a triangular shape, shown in figure 6.9. If the flow is turbulent, the distribution will be broadened: measurement of this broadening is one way of determining the amount of turbulence in the flow.

LDV. As explained previously, the beams in an LDV experiment are crossed over so that an interference fringe pattern is formed in the intersection region, as shown in figure 6.7. The parameters d, d_m, l_m and δ are determined by the laser wavelength λ, the focal length of the lens f, the laser beam dimension D, and the intersection angle θ. For a typical case, with $\lambda = 500$ nm, $\phi = 200$–600 mm, $D = 1$ mm and $\theta = 2$–$15°$, these parameters are $d = 0.12$–0.38 mm, $d_m = 0.1$–0.4 mm, $l_m = 1$–5 mm and $\delta = 2$–15 μm.

If there is a frequency difference Δf between the two beams, the interference pattern generated by the beams is dynamic rather than static. In this case, signals are generated by the difference of the moving fringe pattern and the flow velocity. If dust enters this region, the direction of the flow, as well as its velocity, can be determined. In such experiments, Δf usually is chosen to be in the range of 40–80 MHz.

Comparison of the two techniques. Both L2F and LDV have advantages and disadvantages, and each is appropriate in different circumstances. Typical cases of the beam arrangement for L2F and the interference pattern for LDV are shown in figure 6.10. It can be seen that the distance between the beams in L2F

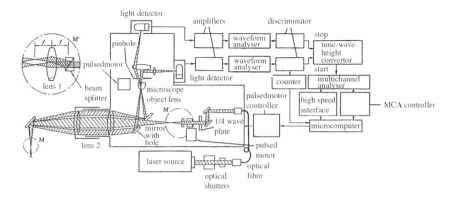

Figure 6.11. Schematic diagram showing a typical experimental arrangement for an L2F measurement system. (Courtesy of Professor H Hayami, Kyushu University.)

is much larger than the fringe size for LDV and so the flow direction can be determined more accurately by the L2F method. In LDV, however, the flow velocity in the direction perpendicular to the fringes is measured.

6.2.2 Examples of Measurements

Flow velocity measurements require lasers that are continuous (cw), have a sufficiently large output power, and, particularly for LDV, are linearly polarized to a high degree. The lasers that best meet these requirements are helium–neon lasers and argon–ion lasers. When a very compact apparatus is desired, diode lasers sometimes can be suitable.

L2F. An example of an L2F instrument is shown in figure 6.11. The main components are a laser, optical elements to direct the beam, a detection system for the Mie scattering, and a signal processing system. In this example, the laser is an argon-ion laser (λ = 488 nm, power = 200 mW) and Mie scattering is observed in the backscattering arrangement. The beam is split into two parts, as shown in the enlarged part of the figure M', then directed to the central part of lens L_2, which focuses the beam into the measurement region M. The two scattered beams pass through the edges of lens L_2, then through the object lens of a microscope and pinholes before being detected. The pinholes are used to both reduce stray light and define the spatial resolution. The beam splitter, shown in M', is rotatable so that the direction of one laser beam can be rotated with respect to the other fixed laser beam. In this way, it is possible to detect the flow velocity in different directions. When this rotation is performed, the pinhole also is rotated. The raw scattered signals vary according to the dust size and the flow velocity. The signal processing system is arranged so that occurrence distribution functions of the type shown in figure 6.8 can be determined from these raw signals.

Figure 6.12. Photograph of the central fan blade in a centrifugal turbo-compressor, showing its complicated structure [1].

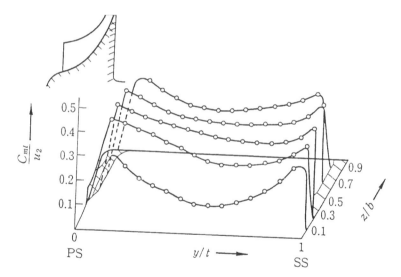

Figure 6.13. Examples of a velocity distribution measured in a centrifugal turbo-compressor using L2F [1].

This kind of measurement system was used to measure the flow velocity inside the centrifugal turbo-compressor shown in figure 6.12 [1]. The impeller of the compressor has diameter of 400 mm, exit width of 35.5 mm, exit angle of 60°, and rotation speed of 22 363 rpm. It has a complex shape and produces complicated flows. An example of the results of that measurement is shown in figure 6.13. This is a two-dimensional plot showing the spatial distribution of

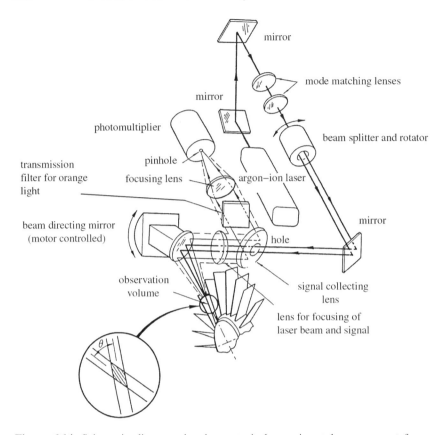

Figure 6.14. Schematic diagram showing a typical experimental arrangement for an LDV measurement system [2].

velocity at the exit of the impeller as functions of the two coordinates y and z. The y and z coordinates are the spatial coordinates in the direction along the blade and height respectively, t is the blade pitch width and b is the blade height. These detailed measurements were compared with simulations of gas flow in this device.

In addition to measurement of average velocity, described above, it also is possible to measure the time-varying velocity. Three-dimensional L2F methods currently are being developed in order to measure three-dimensional flows.

LDV. Figure 6.14 shows an example of an LDV measurement system [2]. The apparatus consists of the same four basic elements as the L2F system described earlier, with two important differences. In this case, the two laser beams are crossed over in the measurement region so that interference fringes are formed. In addition, the Mie scattering from dust particles passing through the interference fringes is collected by a single detector. As mentioned above, the direction in which the flow velocity is measured could be changed by rotating the beam splitter that produces the two beams.

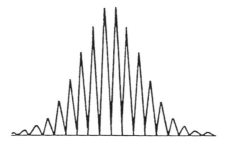

Figure 6.15. Example of the raw signal measured by a LDV apparatus [2].

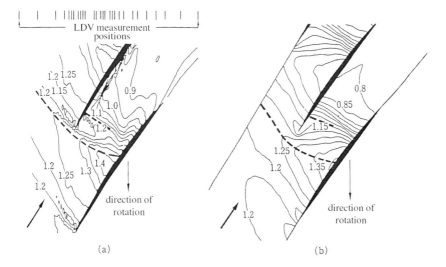

Figure 6.16. Velocity distributions measured in the interior of a transonic axial flow fan [3]. The velocity is shown in units of Mach numbers. (*a*) shows experimental data measured by LDV while (*b*) shows the results of a numerical simulation.

Figure 6.15 is an example of the Mie scattering signal recorded when dust passed through the measurement region. For the determination of flow velocity inside this device, several problems had to be considered. Firstly, the device rotated, causing the blades to pass through the observation region periodically. This interrupted the signal counting and affected the dust distribution and the Mie scattering signal. For this reason, a data processing system was employed to overcome the problem of uneven data acquisition rates. In addition, the dust density was controlled carefully to ensure that only one dust particle at a time passed through the interference fringes.

Figure 6.16(*a*) shows the spatial velocity distribution measured in between the blades of the fan [3]. This type of device is called an *axial flow fan,* because the flow is in the direction of the axis of the device, and the compression ratio is small. In this case, the flow velocity in the regions between the fan blades is

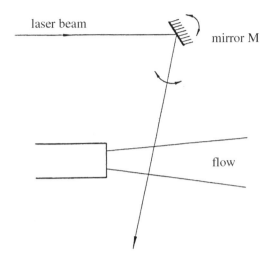

Figure 6.17. Schematic diagram showing a typical experimental arrangement for visualization of a gas flow using LIF.

near the speed of sound, and so this particular device is called a *transonic axial flow fan*. The fan had outer diameter of 513 mm, 22 blades, and the ratio of the radius of the inner cylinder, to which the blades were attached, and the radius of the outer cylinder (called the hub radius) was 0.375. The contour lines in the graph indicate the velocity distributions shown in units of Mach numbers. Figure 6.16(*b*) shows the flow velocity in the device as calculated by a numerical simulation. It can be seen that the broad features of the calculated flow are consistent with the experimentally obtained results.

It can be seen that laser diagnostic methods can be used to measure inside real flow machines, and provide experimental information about the complicated flows in these machines. For this reason, they have become an important part of gas flow research.

6.3 Imaging of Gas Flows by Laser-Induced Fluorescence

The basic principle of laser-induced fluorescence (LIF) is discussed in detail in section 2.2.4 and can be summarized briefly as follows. Laser photons are used to excite atoms or molecules from a low-energy state to a higher-energy state. Absorption only occurs when the photon energy exactly matches the energy difference between particular energy states in that particle. This absorption is detected by observing the fluorescence that occurs when the excited particles lose energy and decay radiatively to lower levels. Measurement of the absolute fluorescence intensity can be used to determine the particle density. Measurement of either the broadening of the absorption profile caused by thermal motion of the particles (the Doppler profile) or the population distribution among various levels can be used to determine the particle temperature. In addition, because electric and magnetic fields affect energy levels, these quantities also can be determined. Electric fields can be determined by measuring Stark profiles and magnetic fields can be determined by measuring Zeeman profiles. These characteristics make LIF extremely useful for plasma measurements, and examples of these types of measurements were discussed in detail in chapter 4. In addition to these plasma applications, an attempt to measure electric fields under thunderclouds using the LIF technique has been reported [4].

Recently, there have been several attempts to use LIF for visualization of gas flows. In these experiments, the fluorescence intensity was recorded directly, either on photographic films or electronically, and the density distribution was determined from the measured distribution of fluorescence intensity. In addition, by measuring the fluorescence from excitation of transitions from different lower levels and thereby determining the density distribution among these levels, the atomic or molecular temperature could be obtained.

Figure 6.17 shows a schematic diagram of such an experiment. A laser beam with wavelength corresponding to an absorption wavelength of atoms or molecules in the gas flow is directed across the flow region. Particles along the entire path of the laser beam will absorb laser light and subsequently fluoresce. If the laser intensity is less than the saturation intensity for the absorption transition, the distribution of fluorescence is equivalent to the density distribution. Two-dimensional images of the fluorescence can be obtained by rotating the mirror and scanning the beam across the entire cross-section of the flow. Alternatively, a two-dimensional image can be obtained without scanning the beam by forming the laser beam into a sheet beam before it is directed across the flow, and observing fluorescence from all parts of the sheet beam simultaneously.

The target of most gas flow studies is flows in air. However, visualization of such flows using species such as oxygen, nitrogen and argon requires very short-wavelength tunable laser beams, and available tunable lasers cannot generate sufficiently large light intensity at the necessary wavelengths. To overcome this problem, LIF usually is observed from small amounts of another species added to the flow for measurement purposes, often referred to as a *tracer*. The tracer species has transitions with wavelengths in the range of tunable lasers such as dye lasers. Alkali metals are possible candidates for this tracer species, but both corrosion of the walls of the flow tube and safety are problems. The most widely used species is iodine (I_2), which, although corrosive, is relatively safe and easy to handle. Figure 6.18 shows the energy band structure of I_2, and table 6.1 shows the absorption wavelengths used for LIF imaging experiments. As can be seen in figure 6.18, there are many possible transitions from the many rotational states: these transitions are shown in the table labelled in the standard way. P and R transitions are transitions in which the rotational quantum number of the molecule changes by $+1$ and -1 respectively.

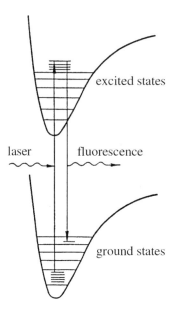

Figure 6.18. Energy levels of the I_2 molecule, showing general excitation and fluorescence transitions used for LIF.

Table 6.1. Transitions of the I_2 molecule used for LIF imaging experiments.

Wavelength (cm^{-1})	Transition	
19 429.5999	P48	(44,0)
	P103	(49,0)
19 429.7317	R98	(58,1)
19 429.8137	P13	(43,0)
	R15	(43,0)
19 429.9277	P46	(50,1)
	P88	(47,0)
19 429.9932	R28	(49,1)

6.3.1 Measurement of Density Distributions

Visualization of gas flows using LIF methods first was performed for flows in which gas expanded rapidly into low-pressure chambers, and for gas flows onto substrates used in thermal CVD processes. Recently, LIF methods have been used to study complicated gas flows in devices such as that discussed in section 6.2. In this section, measurement of the supersonic gas flow from a nozzle into a chamber is used as an example of a density measurement. The flow is complicated, but clear and rather beautiful flow patterns are formed. In this

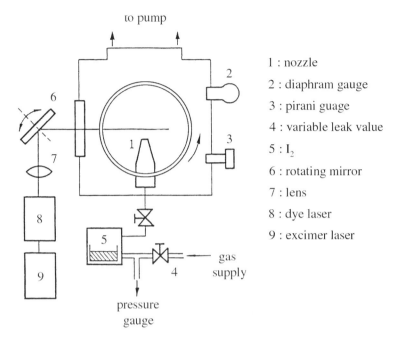

Figure 6.19. Apparatus used for an LIF visualization experiment of supersonic gas flow from a nozzle [5].

experiment, two-dimensional distributions of temperature as well as density were obtained, as will be discussed in the following section.

The experimental apparatus is shown in figure 6.19 [5]. The apparatus consists of three main parts: a device to generate the flow, a laser and optical system for generating an image of the LIF signal, and a detector to record the image. The working gas was argon, which was introduced via a valve into a mixing chamber shown as 5 in the figure. In this mixing chamber, the argon gas was combined with iodine at a pressure of 40–200 Torr before being injected into the main chamber through the nozzle shown as 1 in the figure. The chamber pressure was maintained at a lower pressure, in the range of 1 mTorr–1 Torr. The pressure was monitored using gauges at the walls. The beam from excimer laser pumped dye laser was focused by a lens, shown as 7 in the figure, and directed into the chamber by a mirror. The laser beam could be scanned across a plane in the direction of the gas flow by rotating this mirror. The detection system was an ordinary camera, which had a narrow-band optical filter placed in front of it so that only light at the fluorescence wavelength was transmitted to the detector. The camera shutter was opened for the time it took to scan the beam through the flow, enabling a two-dimensional image of the fluorescence to be obtained.

In Figure 6.20, (*a*) shows an example of a measured image while (*b*) shows the main parts of the flow in a schematic diagram. By comparing the photograph

226   *Measurements in Gas Flow Systems*

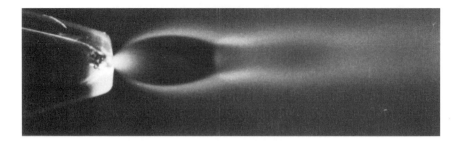

(a)

(b)

Figure 6.20. (*a*) Photographic image of the flow from a jet nozzle, obtained in a LIF imaging experiment. (*b*) Main features of the flow.

in (*a*) with the flow pattern in (*b*), several important points about the flow can be understood. Firstly, it can be seen that the gas rapidly expands from the nozzle mouth, which is the narrowest part of the nozzle. Furthermore, the flow velocity at the nozzle is equal to the speed of sound while the flow in the downstream area, on the right-hand side of the figure, is supersonic. Secondly, the gas expands until the point at which a normal shock wave forms. Thirdly, it is clearly visible that the flow is separated from the background gas in the chamber by a jet boundary. In addition to these direct observations, it is possible to determine a relative density distribution from the distribution of measured light intensity.

6.3.2 Measurement of Temperature Distributions

As can be seen from figure 6.18, a molecule such as I_2 has many different excitation transitions. The fluorescence intensity for each of these transitions can be observed by changing the wavelength of the excitation laser so that transitions from different lower levels are excited selectively. The fluorescence

Imaging of Gas Flows by Laser-Induced Fluorescence

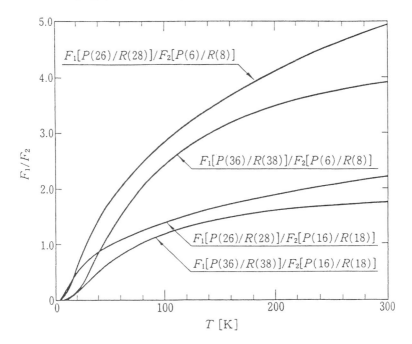

Figure 6.21. Relationship between molecular temperature and the ratio F_1/F_2 of all fluorescence signals obtained using two different excitation wavelengths [5].

intensities measured from the different transitions depend on the densities in each lower level, and so the distribution of intensities provides information about the density distribution among the lower levels. Thus, measurement of the fluorescence on several different transitions can be used to determine the particle temperature. There are many possible fluorescence transitions for each excitation transition, and in the simplest case, the ratio of the fluorescence intensities F_1 and F_2 generated by excitation on only two different transitions can be used to determine the particle temperature. If it is assumed that the density is distributed among the lower levels according to a Boltzmann distribution, the temperature can be determined using a graph such as that shown in figure 6.21 [5].

Figure 6.22 shows fluorescence observed from the flow discussed in section 6.3.1, measured for two different excitation wavelengths. By calculating the intensity ratio from these images, the temperature can be determined directly from figure 6.21. Figure 6.23 shows the temperature distribution along the axis of the flow, determined in this way. It can be seen that the gas cooled rapidly as it expanded away from the nozzle. It can also be seen that up until the onset of the normal shock wave, which is indicated by the temperature jump at $X/D = 7$, the measured cooling of the gas is in excellent agreement with the results of an isoentropy calculation, shown in the figure by the dashed line.

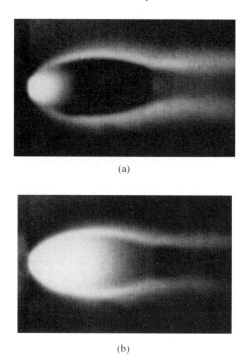

Figure 6.22. Fluorescence intensity distributions measured using different excitation wavelengths. (*a*) shows results for P(16)/R(18) excitation/fluorescence and (*b*) shows results for P(26)/R(28) excitation/fluorescence.

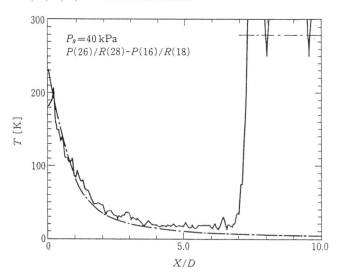

Figure 6.23. Axial temperature distribution determined using the ratio of the fluorescence signals shown in figure 6.22. The dashed line is the temperature distribution predicted by an isoentropy calculation.

It should be noted that the above discussion ignores the effects of collisions on the upper state densities of the excited I_2 molecules. This was not a problem for this particular experiment because the gas density and temperature were low, but at higher pressures and/or higher temperatures, this effect may have to be considered.

REFERENCES

[6.1] Krain H 1988 *ASME J. Turbomachinery* **110** 122
[6.2] Powell J A, Strazisar A J and Seaholz R G 1981 *ASME J. Eng. for Power* **103** 424
[6.3] Rierzga M J and Wood J R 1985 *ASME J. Eng. for Power* **107** 436
[6.4] Gavrilenko V P, Muraoka K and Maeda M *Japan. J. Appl. Phys.* (submitted)
[6.5] Niimi T, Fujimoto T and Shimizu N 1990 *Opt. Lett.* **15** 918

Chapter 7

Laser Processing Measurements

7.1 Laser Processing

In this chapter, the term 'laser processing' will be used in a broad sense to mean any type of process in which the properties of a material are changed, physically or chemically, when the material is irradiated by laser radiation. For high-power lasers, this is the most important application. Some types of laser processing, such as thermal processing and machining using CO_2 lasers and Nd:YAG lasers, are already in widespread use in industry [1].

The spatial and temporal properties of laser light can be controlled by changing the pulse width and the beam diameter, and this enables generation of

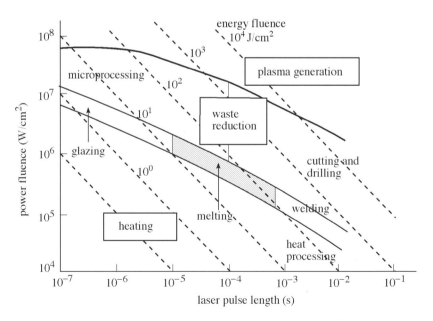

Figure 7.1. Diagram showing the thermal effects in laser irradiation and their applications.

beams with fluence that varies over an extremely wide range. Figure 7.1 shows the variety of processes that are possible by utilizing this wide range of fluence, including controlled heating, melting, vaporization and ionization. Table 7.1 shows industrial applications of laser processing divided into the four areas of cutting, welding, surface treatment and evaporative heating. The first three of these categories are self-explanatory, and the fourth, evaporative heating, is the process in which a pulsed laser is focused onto the surface of a material, producing a high-temperature vapour or plasma called a plume, which is used for thin-film deposition or for the generation of small particles.

Nd:YAG lasers are well suited for industrial fine processing and machining, because the laser light can be transmitted by optical fibres and then focused to a small point. Excimer lasers are used in a different way, because the photon energy of ultraviolet excimer laser beams is the same order as the binding energy of many molecules, and so the laser photons are capable of dissociating these molecules [2,3]. Similar dissociation can be achieved via multiphoton absorption if a pulsed laser with very high instantaneous power is used, and even TEA CO_2 lasers, which operate at mid-infrared wavelengths, can cause dissociation. However, much better control of the production of chemical species is possible with excimer lasers, and the development of powerful excimer laser sources has allowed rapid development of the field of the laser chemistry. Although it is true that thermal reactions and photochemical reactions often occur simultaneously, photochemical reactions can be induced at low temperature, which is a great advantage.

Photochemical reactions induced by light from Hg lamps have been used for many years in chemical plants. Much higher fluence, of the order of 10^6–10^8 W cm^{-2}, is achievable by replacing these lamps with excimer lasers, and this allows radicals and excited species to be generated at high densities. In addition, secondary reactions do not occur, the selectivity of the process is improved and the overall efficiency is increased. However, laser processing is not very cost-efficient for the production of bulk materials, because the photon cost of excimer lasers is high. For this reason, laser processing techniques are practical only for fabrication of high added-value products. It is expected that laser processing methods will be used most widely in the area of fabrication of functional devices such as semiconductors.

Table 7.1. Examples of thermal treatment processes that use lasers.

Removal processes	cutting, drilling, marking, precision machining (trimming, scrubbing, mask repairs)
Welding	soldering, sheet-welding, micro-welding, ceramic jointing
Surface treatment	annealing, cladding, glazing, making of metal alloys
Evaporative heating	spraying, vapour deposition, small particle generation

Table 7.2. Comparison of laser photon energies with the dissociation energies of common molecules.

Laser	Wavelength	Photon energy	
CO_2 laser	10.6 μm	0.1 eV	(2.7 kcal mol^{-1})
Nd:YAG laser	1.06 μm	1.0 eV	(27 kcal mol^{-1})
XeF laser	350 nm	3.5 eV	(81.1 kcal mol^{-1})
XeCl laser	308 nm	4.0 eV	(92.2 kcal mol^{-1})
KrF laser	249 nm	5.0 eV	(114.1 kcal mol^{-1})
ArF laser	193 nm	6.4 eV	(147.2 kcal mol^{-1})
F_2 laser	157 nm	7.9 eV	(180.1 kcal mol^{-1})

Bond	H	C	N	O	S
H–	103	81	92	109	84
C–		84	64	76	65
C=		140	149	176	92
C≡		191	222	256	181
N–			36		
N=			110		
N≡			124		

bond-breaking energy at 25°C (kcal mol^{-1})

Table 7.3. Examples of laser processing applications for device fabrication.

Process	Film material	Process gas
Laser CVD	Si	SiH_4, Si_2H_6
	SiO_2	$SiH_4 + N_2$
	C	C_2H_2, CH_4
	Al_2O_3	$Al(CH_3)_3 + N_2O$
	W	$W(CO)_6$, WF_6
Etching	Si	Cl_2
	SiO_2	$NF_3 + H_2$
	GaAs	CCl_4, HCl
Doping	B in Si	BCl_3, $B(CH_3)_3$
	Al in Si	$Al(CH_3)_3$
	Si in GaAs	SiH_4
Epitaxial growth	GaAs	$Ga + AsCl_2 + H_2$
Ablation (PLD)	polymers, metals	
	YBCO, dielectrics	O_2

Some examples of laser processing techniques based on excimer lasers are shown in table 7.3. In *chemical vapour deposition* (CVD), a laser is used to dissociate molecules in a gas, and a thin film is grown on a nearby substrate. One example of this is the production of Si films using Si_2H_6 gas, which has strong absorption bands at the ArF excimer laser wavelength. It also is possible to use SiH_4 gas, because although SiH_4 does not absorb well at ArF excimer laser wavelengths, multiphoton absorption causes the necessary dissociation.

In the process called *photo-etching*, reactive species such as Cl and F are produced by photo-dissociation of parent gases such as Cl_2, HCl and NF_3. This kind of process also can be used for photo-chemical cleaning and oxidizing of wafer surfaces. In the process called *doping*, individual sections of a substrate surface are irradiated with light, allowing specific elements from the dopant gas to be incorporated into these areas on the surface. Extremely shallow doping is possible by using excimer lasers, and doping depths of the order of 10 nm depth are achievable. The process called *ablation* can be performed as a thermally activated process, but thermally induced strain can be avoided by using excimer lasers, and, because chemical processes predominate, even materials such as polymers can be cut. In addition, if a substrate is placed close to the ablation region, thin films can be deposited. This process is called *pulsed laser deposition* (PLD), and the material from the target can be directly transferred to the film. Thin-film fabrication by the PLD method is the most promising technique in laser processing. The apparatus is simple and a variety of thin films, from polymers to inorganic crystals, can be fabricated by changing the target material. The most successful application of this method is the production of high-temperature superconducting thin films.

In addition to the applications described above, excimer lasers are the focus of research efforts aimed at *micro-lithography*. A device used for photolithography, called a *stepper*, is shown schematically in figure 7.2. The stepper is a type of projector that is used to project a reduced mask pattern onto the wafer. The exposed areas on the wafer then can be selectively etched. The narrowest dimension that can be projected determines the degree of integration that can be achieved on the integrated circuit. This dimension, R, is given by

$$R = \frac{k\lambda}{NA} \quad (7.1)$$

where λ is the wavelength of the light source, k is a constant of the order of 0.5 and NA is the numerical aperture of the lens system. For lithography, Hg lamps have been used as the light source, with the g line (λ = 436 nm) and the i line (λ = 365 nm) being used. By replacing these lamp sources with excimer lasers, the resolution is improved due to the shorter wavelength, and the processing quality is improved because of the monochromatic nature of the laser light, which greatly reduces chromatic aberration. If we consider a system with k = 0.5 and NA = 0.5, then for a KrF laser (λ = 248 nm), $R \sim$ 0.25 mm, and for an ArF laser (λ = 193 nm), $R \sim$ 0.2 µm. The R value for a KrF laser is sufficient for fabrication of DRAM with capacity of 256 Mbytes to 1 Gbytes.

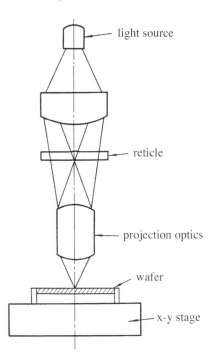

Figure 7.2. Optical arrangement of stepper device used in photo-lithography.

Isotope separation is another material processing area that can be performed effectively using laser processing techniques. In this process, the U^{235} isotope is separated from uranium ore, which contains only 0.7% of this isotope and mainly consists of U^{238}. Figure 7.3 shows an isotope separation system that currently is proceeding from the research stage to actual industrial use [4]. In the first step, uranium ore is placed in a vacuum system and evaporated with an electron beam. The U atoms then are excited using a dye laser with an oscillator–amplifier system pumped by a Cu-vapour laser that has average power of 100–200 W and is operated at repetition rates of 5 kHz or higher. Four different wavelengths of the dye laser are used to selectively excite and ionize the U^{235} atoms, and an electric field is used to separate the ions. This scheme is called the *atomic method* of separation. An alternative technique, called the *molecular method*, also is being developed. In this method, UF_6 molecules are dissociated selectively by an infrared laser.

7.2 Measurement Methods in Laser Processing

7.2.1 Different Methods and their Advantages

Among the many kinds of laser-aided measurement techniques discussed in this book, laser spectroscopic techniques for gas phase measurements are especially

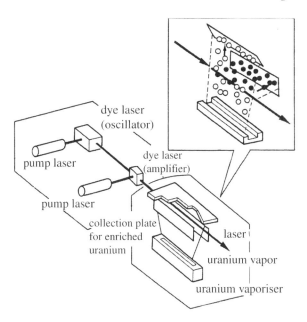

Figure 7.3. Schematic diagram of a uranium isotope separator based on the atomic method [4].

important for laser processing. Table 7.4 contains a list of the main techniques used in this field. Passive emission measurements can be used to obtain information about the properties of the atoms and molecules in the process. This can be considered a type of laser-induced emission spectroscopy, in which the processing laser acts as the excitation source.

When laser-aided measurement systems are applied to laser processing, a separate laser is used to probe the processing region, as shown in figure 7.4. Phenomena such as transmission, scattering, fluorescence, ionization and sound generation then are used to obtain information about the processing conditions. These kinds of active measurement methods provide much greater detection sensitivity, and greater temporal and spatial resolution than the passive emission techniques mentioned above. The species of the scattering particle cannot be identified when Rayleigh and Mie scattering methods are used, but spectroscopic methods can be used to selectively detect different species of atoms or molecules, and so properties such as density can be measured. In emission measurements, only the density of *excited* states can be determined, but when laser methods are used, excitation from any state including the ground state is possible and the density in any state can be determined. In general, the size of the signal is proportional to the density, and these methods are used primarily as density measurement techniques.

It is possible to measure temperature as well as density with laser-aided diagnostic techniques. The particle temperature can be determined by measuring

Table 7.4. Laser-aided measurement methods used for laser processing in gases.

	Measurement technique	Measurement objective	Abbreviation
Methods based on fixed-wavelength lasers	Rayleigh scattering	molecular density, temperature	
	Mie scattering	density, size and velocity distribution of small particles	
	Raman spectroscopy	molecular density, temperature	
	emission spectroscopy	atomic/molecular density temperature	
	laser mass spectroscopy	atomic/molecular/cluster density	LMS
Methods based on tunable lasers	absorption spectroscopy • infrared absorption spectroscopy • intra-cavity absorption spectroscopy	molecular density molecular density	AS IRAS ICAS
	fluorescence spectroscopy	atomic/molecular density velocity distribution	LIF
	resonant ionization spectroscopy • multi-photon ionization • optogalvanic spectroscopy	atomic/molecular density atomic/molecular density	RIS REMPI OGS
	opto-acoustic spectroscopy	atomic/molecular density	OAS
	nonlinear spectroscopy • coherent anti-Stokes Raman spectroscopy • degenerate four-wave mixing	molecular density, temperature atomic/molecular density	CARS DFWM

the *Doppler profile* of a spectral line, as this depends on the species' velocity distribution. For the case of molecules, the distribution of particles amongst the various rotational energy levels depends on temperature, and this rotational temperature can be determined by fitting a theoretical curve to an experimentally measured distribution. This method has been developed into a standard temperature-measuring instrument, based on the CARS technique. As will be explained later, the *time-of-flight* (TOF) method also can be used to determine particle velocity, because in this method the particles are generated instantaneously by an optical process triggered by a pulsed laser. TOF

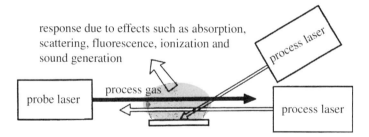

Figure 7.4. Schematic diagram of a laser-aided measurement system for laser processing.

techniques do not require the highly monochromatic light source that is necessary for Doppler profile measurements and hence are easier to apply.

The main merit of using laser spectroscopy in laser processing environments is that particular species can be detected selectively, and that species' density and velocity can be determined *in situ*. This enables quantities such as particle diffusion constants and reaction rate constants to be determined, and hence, information about the dynamics of the process to be obtained.

The properties of the film that is produced in these laser processing applications can be determined by a variety of techniques, including x-ray diffraction, Raman spectroscopy, photoelectron spectroscopy, light interferometry and ellipsometry. These diagnostic techniques for solid-sate materials, however, are not discussed in this book. The methods outlined in this section can be used to make highly sensitive *gas phase* measurements with good spatial resolution very close to the material surface. In this way, information about the interaction between the gas phase species and the surface can be obtained.

7.2.2 Detection of Atomic and Molecular Species

In laser processing, an extremely wide range of species have to be measured because radical species with short lifetimes are important, as well as stable species such as molecules and atoms. Table 7.5 shows a list of atoms, molecules and radicals important in CVD processes involving carbon, silicon and germanium, and the measurement methods used to detect them [5]. Table 7.6 shows a list of light elements such as H, C, N and O, and halogen atoms that are important in laser etching, together with methods used to detect them. This table also includes species that are not related directly to laser processing. Abbreviations listed in the table are explained in table 7.4.

As can be seen from these tables, LIF is the most widely used method. However, LIF cannot be used when the particles in the excited state are quenched by collisions. Species with long-lived excited states are susceptible to quenching, and the effects become more severe as the gas pressure becomes

Table 7.5. Examples of measurements of radical species containing C, Si and Ge. (Original references are cited in [7].

Particle	Transition	Excitation wavelength (nm)	Method	Particle	Transition	Excitation wavelength (nm)	Method
C_2	$a^3\Pi - d^3\Pi$	516	LIF	Si	$3p^3P - 4s^3P$	251	LIF
			2D LIF		$3p^3P - 4s^3P$	251	AS
	$X^1\Sigma - A^1\Pi$	691	LIF		$3p^3P - 4s^3P'$	408	RIS
C_3	$X^1\Sigma - A^1\Pi$	405	LIF	Si_2	$X^3\Sigma - H^3\Sigma$	392	LIF
		405	SEP	SiH	$X^2\Pi - A^2\Delta$	413	LIF
CH	$X^2\Sigma - A^2\Delta$	431	LIF	SiH_2	$X^1A_1 - A^1B_1$	610	LIF
			2D LIF			510	RIS
CH_2	$a^1A_1 - b^1B_1$	590	LIF		ν_I-band		CARS
	ν_2-band		IRAS		$X^1A_1 - A^1B_1$		ICAS
			SEP		$X^1A_1 - A^1B_1$	570	AS
CH_3	$2p^2A'' - 3p^2A''$	333	RIS	SiH_3	$X^2A_1 - E^2A_2''$	390	RIS
			2D RIS		ν_2-system		IRAS
	$2p^2A'' - 3p^2A''$		OGS	SiO	$X^1\Sigma - A^1\Pi$	221	LIF
	ν_2-band		IRAS	SiN	$X^2\Sigma - B^2\Sigma$	396	LIF
	ν_1-band		CARS	SiF	$X^2\Pi - A^2\Sigma$	437	LIF
CF	$X^2\Pi - A^2\Delta$	224	LIF	SiF_2	$^1A_1 - ^1B_1$	223	LIF
CF_2	$X^1A_1 - A^1B_1$	234	LIF	Ge	$3p^3P - 4s^3P$	265	LIF
CCl	$X^2\Pi - A^2\Delta$	278	LIF	GeH	$X^2\Pi - A^2\Delta$	360	LIF
CCl_2	$X^1A_1 - A^2B_1$	515	LIF	GeH_3	$X^2A_1 - ^2A_2''$	419	RIS

Note: SEP is stimulated emission pumping and 2D represents a two-dimensional measurement. The other measurement method abbreviations are given in table 7.4.

higher. Figure 7.5 shows the relationship between fluorescence lifetime, quenching rate and the fluorescence efficiency. Particles in the lower left part of the graph are well suited for LIF methods while those in the top right are difficult to detect with this method. The important species in CVD processes include radicals such as CH_n, SiH_n and GeH_n. These species can be detected by LIF methods if $n \leq 2$. RIS or absorption methods have to be used for species such as CH_3 and SiH_3.

Another important factor in determining the feasibility of laser-aided diagnostic methods is the wavelength of the transition that is used. Most of the species shown in table 7.5 can be detected using tunable dye lasers as the laser source. These sources cover the wavelength region from 200 nm to about 1 μm. Most of the molecular species can be detected using infrared absorption spectroscopy (IRAS) provided a suitable IR laser exists.

Most of the low mass-number atomic species shown in table 7.6 have resonant transitions from the ground state with wavelengths in the deep vacuum ultraviolet, out of the range of conventional dye lasers. Detection of these

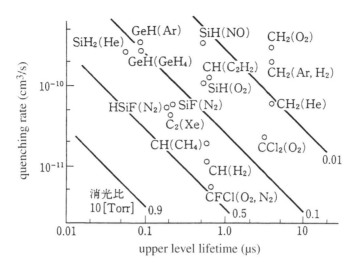

Figure 7.5. Relationship between fluorescence lifetime and quenching rate for different species [5]. The solid lines indicate the quenching ratio while the ambient gas for each case is shown in parentheses.

Table 7.6. Measurement of light elements and halogens. (Original references are cited in [7].

Particle	Transition	Excitation wavelength (nm)	Method	Particle	Transition	Excitation wavelength (nm)	Method
H	1s – 2s	121.6	LIF	O	$2p^3P – 3p^2P$	266 (2P)	LIF
	1s – 3s, 3d	205 (2P)	LIF				1D LIF
	1s – 3p	243 + 656 (2P+1)	2D LIF				TPSE
	1s – 3s, 3d	205 (2P)	TPSE			226 (2P + 1)	OGS
	1s – 2s	121.6	AS	F	$^2P^0 – {}^2D^0$	170 (2P)	LIF
	1s – 2s	224 + 266 (2P + 1)	OGS		$^2P_{3/2} – {}^2P_{1/2}$		IRAS
	1s – 2s	364 (3P + 1)	RIS		$^4P_{3/2} – {}^4D_{5/2}$	690	LIF
C	$^3P – {}^3D$	287 (2P)	LIF	Cl	$^2P – {}^2F$	210 (2P)	LIF
	$2p^3P – 3p^3P$	280 (2P)	TPSE		$^2P – {}^4S$	233 (2P)	LIF
	$2p^3P – 3s^3P$	166	LIF		$^2P – {}^4S$	233 (2P)	TPSE
	$2p^1D – 3s^1P$	193	LIF		$^2P – {}^4S$	233 (2P)	RIS
N	$2p^4S – 3p^4P$	211 (2P)	LIF		$^2P_{3/2} – {}^2P_{1/2}$		IRAS

Note: TPSE is two-photon excited stimulated emission pumping and 2P represents photon excitation. The other measurement method abbreviations are given in table 7.4.

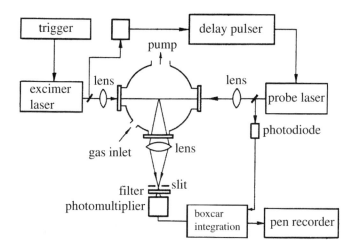

Figure 7.6. Schematic diagram of an LIF measurement system for laser CVD.

species requires one of the following three approaches: use of a special laser source that is tunable in the VUV region; use of high-power UV or visible lasers with multiphoton absorption; or use of transitions from excited states.

The first approach is widely applicable in principle, but requires the construction of a special laser source that uses nonlinear conversion in gases to generate the VUV beam, and it usually is difficult to construct such a laser [6]. For this reason, there are few examples of measurements actually made with such systems, although table 7.6 does contain one example, a measurement of hydrogen using the Lyman α transition. The third approach can only be used in environments such as plasmas and flames in which there is a sufficiently high density of excited species. The second approach, multiphoton excitation, is by far the most widely used method of detecting these species. The laser source, usually a dye laser pumped by a Nd:YAG or an excimer laser with very high instantaneous power, is relatively simple. However, the detection sensitivity for multiphoton excitation is poorer than that for single-photon excitation, and the detection limit depends strongly on the power of the laser source.

Another factor that should be considered in multiphoton excitation experiments is that amplified spontaneous emission (ASE) can occur if there is a population inversion between the upper and lower levels of the fluorescence transition. When this occurs, the LIF fluorescence intensity will not depend linearly on the species density. There are even examples of measurements of density made using ASE.

7.3 Examples of Laser Processing Measurements

There are many different examples of laser spectroscopic techniques applied to laser processing. The following discussion mainly contains examples of measurements made during CVD and ablation research performed in the authors' own laboratories. The examples used below involve the measurement of particle densities of gas-phase species. These laser-aided measurements, made with high temporal and spatial resolution, provided understanding of the dynamic nature of each type of process.

7.3.1 Measurements of Laser CVD Processes

In this section, we will discuss LIF and RIS measurements of radicals generated by laser-induced dissociation of gases containing carbon and silicon [7–9].

LIF measurements. Figure 7.6 shows the experimental apparatus used for this measurement [8]. The excimer laser used for dissociation was an ArF (λ = 193 nm) excimer laser. The carbon-based process gases were acetylene (C_2H_2), and methanol (CH_3OH), and the silicon-based gases were silane (SiH_4) or disilane (Si_2H_6). He or Ar was used as a buffer gas. The probe laser, an excimer laser pumped dye laser, was triggered after the ablation laser at a time determined by a delay pulser. The time dependence of the radical density could be measured by varying this delay time. The dye laser had output energy of 2–3 mJ and spectral width of about 90 GHz. Although absorption spectroscopy requires spectrally very narrow laser sources, this was not necessary for this experiment. An f = 250 mm lens was used as the collection lens for the fluorescence.

The LIF signal was observed in a direction perpendicular to the laser beam and detected by a photomultiplier. An optical filter was placed in front of the detector in order to reduce the effect of emission associated with the dissociation, and a viewing dump consisting of a set of knife-edges was used to reduce the amount of stray light. The photomultiplier signal was averaged using a boxcar integrator and then analysed using a microcomputer.

Table 7.7 shows the carbon- and silicon-based radicals that can be measured during CVD, and the excitation wavelength for each species. In the

Table 7.7. Transitions used for LIF measurements of carbon- and silicon-containing radicals in CVD processes.

Radical	Transition	Excitation wavelength (nm)
C_2	$A^1\Pi_u - X^1\Sigma_g^+$	691
C_2	$d^3\Pi_g - a^3\Pi_u$	516
CH	$A^2\Delta - X^2\Pi$	431
CN	$B^2\Sigma^+ - X^2\Sigma$	388
Si_2	$H^2\Sigma^+ - X^3\Sigma$	390
SiH	$A^1\Delta - X^2\Pi$	410
SiN	$B^2\Sigma - X^2\Sigma$	395
SiO	$A^2\Pi - X^1\Sigma$	221

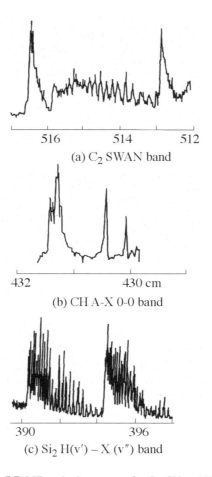

Figure 7.7. LIF excitation spectra for C_2, CH and Si_2 [8].

experiment described here, C_2, CH, Si_2 and SiH were detected. Measured excitation spectra for these species are shown in figure 7.7. In these measurements, the delay time was held constant, and the LIF signal intensity was recorded as the dye laser wavelength was scanned. Although CH_2 and SiH_2 could not be measured in this experiment, these have been detected in other experiments, as listed in table 7.5.

Figure 7.8 shows the signal intensity measured when the laser wavelength was fixed at a value for which the CH signal intensity was maximum, and the ArF laser power then varied [8]. The gradient of the data in the graph is larger than unity, which indicates that two-photon absorption contributed to the generation of CH and C_2. For di-silane, which has absorption bands extending to ArF laser wavelengths, dissociation mainly occurs due to single photon absorption.

Further measurements were made in this experiment. Figure 7.9(*a*) shows

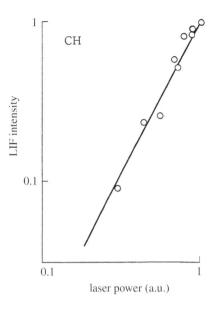

Figure 7.8. Laser power dependence of the density of CH radicals generated by dissociation of methane gas by an ArF excimer laser [8].

the C_2 LIF intensity measured as the delay between the ArF laser and the probe laser was varied [8]. In this experiment, the total pressure, including the Ar buffer gas pressure, was held constant at 20 Torr, and the C_2H_2 partial pressure was varied in the range of 10–100 mTorr. From the graph, it can be seen that the C_2 density decreased exponentially with time, and that the decay rate increased as the C_2H_2 partial pressure increased. This indicates that the loss mechanism for the C_2 ($a^3\Pi_u$) radical was two-body collision with C_2H_2 molecules. The two important reactions are

$$\begin{aligned} C_2 (a^3\Pi_u) + C_2H_2 &\rightarrow 2C_2H \quad \text{or} \\ C_2 (a^3\Pi_u) + C_2H_2 &\rightarrow C_2(X^1\Sigma_g) + C_2H_2. \end{aligned} \quad (7.2)$$

Figure 7.9(b) shows the dependence of the decay rate of C_2 on the C_2H_2 partial pressure, determined from the data in figure 7.9(a). From this data, a rate constant of 6.1×10^{-11} cm³ s⁻¹ for the reactions in equation (7.5) was determined.

The radicals that are generated by the primary dissociation reaction move towards the substrate due to diffusion, but during this process, their density decreases due to secondary reactions either with particles in the background gas or with other radicals. The following equation, which includes both diffusion and the secondary reactions, describes this process:

$$\frac{\partial n}{\partial t} = D^2 n - \sum_i k_i n_i n. \quad (7.3)$$

In this expression, n is the density of the radical of interest, D is its diffusion

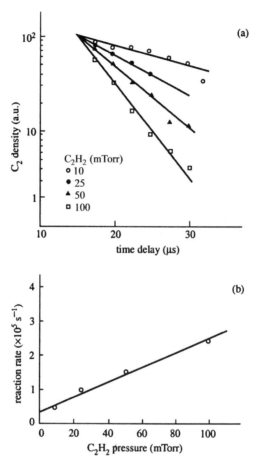

Figure 7.9. (*a*) Density of C_2 radicals generated by dissociation of acetylene (C_2H_2) by an ArF excimer laser, showing the dependence of the C_2 density on the delay time between the ablation laser and the measurement laser [8]. (*b*) Dependence of the reaction rate on the C_2H_2 pressure.

coefficient, and n_i and k_i are the density and rate constant for reactions with the *i*th species. For the experiment described here, the main reactions were those with C_2H_2, as discussed above.

Figure 7.10 shows the spatial distribution of C_2 radicals at a time 20 µs after the dissociation, for the case of (a) Ar and (b) H_2 buffer gas [9]. This distribution was obtained by scanning the probe-laser beam spatially. For both cases, the total pressure was 20 Torr, but the results indicate that diffusion was much faster in the lighter H_2 gas. The dashed line in each figure is a profile fitted to the data based on equation (7.6). From this curve-fitting procedure, diffusion constants for buffer gas pressure of 1 Torr were determined to be $D = 210$ cm² Torr s⁻¹ for Ar and $D = 550$ cm² Torr s⁻¹ for H_2.

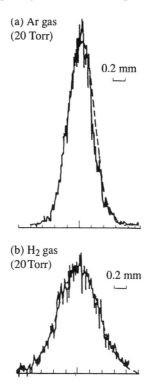

Figure 7.10. Spatial distribution of the C_2 radical density, at a time 20 µs after ablation, obtained by spatially scanning the probe laser [9]. The dashed line is the result of a calculation based on a diffusion equation. The profile in (*a*) is for Ar and (*b*) using H_2 as the buffer gas.

RIS measurements. A typical experimental apparatus for RIS measurements is shown in figure 7.11 [10]. The most noticeable difference from an LIF apparatus is the pair of parallel, planar electrodes inserted into the vacuum chamber in order to measure the electron/ion pairs generated by the probe laser. In this apparatus, the ion current was passed through a pre-amplifier and then averaged using a boxcar integrator.

Radicals such as SiH_3 and CH_3 are believed to play an important role in the formation of thin films containing silicon and carbon, but these species cannot be detected by LIF methods because they do not fluoresce. Figure 7.12 shows an example of a RIS excitation spectra measured when methanol was the parent gas. The sharp peak at 333.4 nm is due to two-photon excitation of CH_3 to the $3p^2A_2''$ level and subsequent ionization. The broad peak in the region of 332.5 nm was observed even when the dissociation laser was not used, and so was considered to be due to either direct ionization of CH_3OH itself or impurities in the gas.

Figure 7.13 shows the RIS spectrum of Si atoms formed by dissociation of

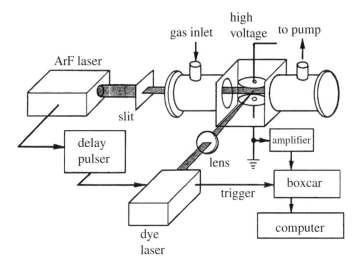

Figure 7.11. Schematic diagram of a RIS measurement system for laser CVD.

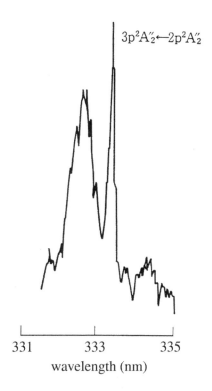

Figure 7.12. RIS spectrum of CH_3 radicals produced from dissociation of methane gas by an ArF laser [10].

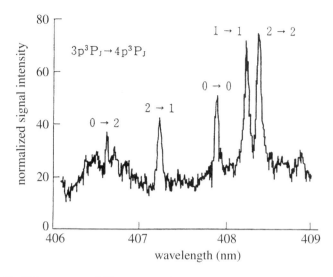

Figure 7.13. RIS spectrum of Si atoms produced by dissociation of di-silane gas by an ArF laser [11]. A 2+1 photo-ionization scheme was used.

di-silane [11]. The peaks in the spectrum are due to ionization via two-photon excitation of $3p^3P_J - 4p^3P_J$ transitions. A similar signal was obtained in the region of 385 nm, but this signal was determined to be due to ionization of the $3p^1D_3$ metastable level of Si. By using both these signals, the species that depend on reactions with ground state and metastable state atoms could be differentiated. A RIS signal at 365 nm due to three-photon excitation of H atoms also was observed.

7.3.2 Measurements of Laser Ablation Processes

In ablation applications that use excimer lasers, chemical processes rather than thermal processes are dominant. This means that much more precise processing is possible. A good example of this precise processing is the drilling of small diameter holes in polymer plates. The technique of thin-film deposition based on laser ablation is called pulsed laser deposition (PLD). In this method, the ablation target is made of the same material as the desired thin film, and the aim of the PLD method is to transfer the material from the target, where it exists in bulk form, to a high-quality crystalline thin film on the substrate. This method has the advantages that the apparatus is simple, the deposition speed is high, and the target material may be chosen freely. Probably the most successful application of PLD is the fabrication of high-temperature superconducting thin films.

In this section, we will discuss laser-aided diagnostics of the most common PLD application, which is fabrication of YBCO superconducting thin films

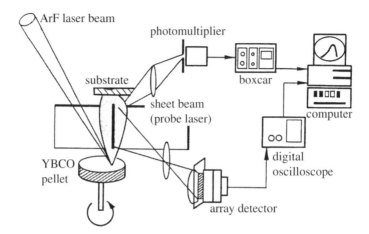

Figure 7.14. LIF-TOF measurement apparatus for detection of particles generated by ablation of a YBCO target using an ArF laser. When the one-dimensional array detector was used, a delay pulser was placed between the ArF laser and the dye laser.

[12,13]. In this field, there are many examples of LIF measurements of atoms and radicals ejected from the target, and of Mie scattering measurements of the density of small particles. Absorption and non-resonant ionization methods also have been used as effective diagnostic techniques. In the discussion here, a single-point LIF measurement will be presented first, and then imaging LIF experiments using one-dimensional and two-dimensional array detectors will be described.

Single-point measurements. The experimental apparatus for this set of measurements is shown in figure 7.14. The target material, which was a ceramic form of $YBaCu_3O_{7-\delta}$ (YBCO), was placed on a rotating mounting stage and irradiated at 45° by an ArF excimer laser beam. The detection systems for both a single-point measurement and a one-dimensional spatial distribution measurement are shown in the figure. For the single-point detection, the beam from a pulsed dye laser was directed parallel to the target surface and the LIF signal was detected from a point directly above the ablation region. The light was collected by a lens, detected by a photomultiplier and averaged using a boxcar integrator. A XeCl laser-pumped dye laser was used as the probe laser, and the time between the ArF ablation laser and the dye laser was variable.

In this experiment, the species listed in table 7.8 were identified by scanning the wavelength of the probe laser and measuring the resulting fluorescence. LIF detection of Cu at $\lambda = 324.8$ nm also is possible, although it was not measured in this experiment. The aim of LIF measurements usually is to determine the particle density, but in the experiment described here, information about the velocity distributions also was desired. One method of gaining information about the particle velocity is to measure the Doppler-

Table 7.8. Excitation wavelengths for particles generated during ablation of YBCO.

Particle	Transition	Excitation wavelength (nm)
Ba	$6p^1P_1^0 - 6s^1S_0$	553.5
Y	$y^2D_{5/2} - a^2D_{5/2}$	412.8
BaO	$A^1\Sigma^+ - X^1\Sigma^+$	535
CuO	$A^2\Sigma - X^2\Pi_{3/2}$	606
YO	$A^2\Pi_{1/2} - X^2\Sigma$	613

broadened profile of spectral lines, but this requires a very monochromatic laser source, in nearly single-mode operation. In the present experiment, in which all the particles of interest were ejected from the target surface at virtually the same time, the TOF was used. By this method, the velocity distribution can be determined relatively straightforwardly using a probe laser with a broad spectral width.

Figure 7.15 shows the velocity distribution of Ba atoms at a position 60 mm above the target surface, measured under vacuum conditions [14]. The horizontal axis is the delay time between the ArF laser and the dye laser. The four graphs show the change in the velocity distribution as the ArF laser fluence

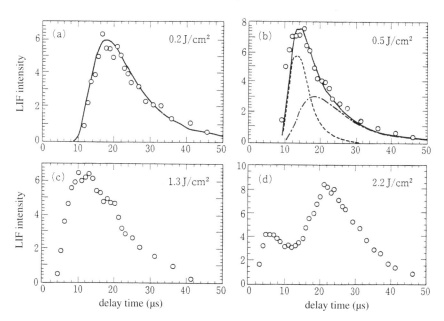

Figure 7.15. LIF-TOF profiles of Ba atoms, measured at a point 60 mm above the target under vacuum conditions [14]. The four graphs show profiles measured for different ArF laser fluence.

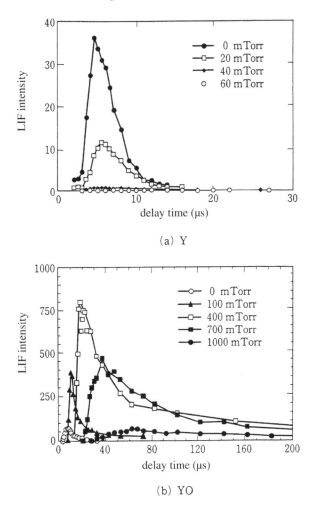

Figure 7.16. LIF-TOF profiles of (a) Y atoms and (b) YO molecules measured at a point 30 mm above the target [15]. The different curves show profiles measured at different values of the O_2 ambient gas pressure.

was varied. It should be noted that the measured range of energies, in the several to 10 eV range, is much greater than can be generated by thermal processes alone. At low fluence, the velocity distribution could be fitted with a shifted Maxwellian distribution, shown by the solid line in the figure. At higher fluence, shown in figure 7.15 (c) and (d), the distribution consists of two components: a low-velocity part and a high-velocity part.

During actual YBCO thin-film fabrication, the ablation is performed in an oxygen atmosphere, at a pressure of 200–500 mTorr. Figure 7.16(a) shows the TOF profile of Y atoms measured in an oxygen gas environment [15]. It can be seen that the Y atom density decreased drastically as the oxygen gas pressure

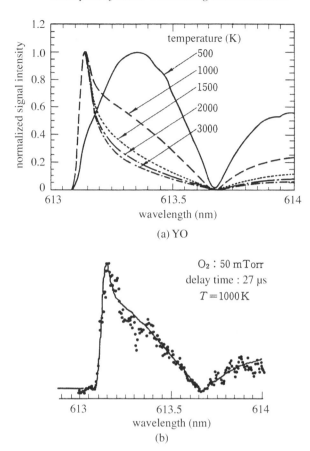

Figure 7.17. Spectrum of a rotational line of the YO molecule [13]. (*a*) shows spectra calculated assuming various temperatures. (*b*) shows an example of a measured spectrum.

was increased. In comparison, the YO radical density, shown in figure 7.16(*b*), increased as the O_2 pressure increased. This indicates that the Y atoms ejected from the target reacted with oxygen in the background gas to form YO. The rate constant for this reaction was determined from this data. From figure 7.16(*b*), it also can be seen that when the oxygen pressure is high, the YO velocity decreased and the density appeared to decrease. This tendency can be attributed to increased diffusion of YO radicals due to O_2 collisions.

Particle temperature can be determined by measuring the rotational spectrum of molecules in the plume, using the fact that the distribution of molecules amongst rotational levels depends on the particle temperature. In this experiment, this measurement was performed using YO molecules. Figure 17.17(*a*) shows calculated excitation spectra for the 0–0 band of the A–X rotational transition in YO molecules, assuming a Boltzmann distribution of particles in the rotational sub-levels [13]. Although the temperature dependence

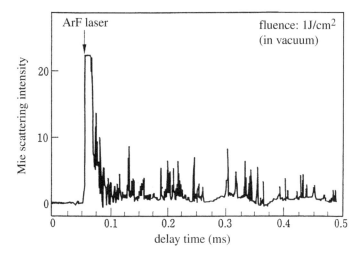

Figure 7.18. Time dependence of the Mie scattering signal of an Ar-ion laser (488 nm) from small particles generated in the ablation process. The initial spike is due to emission from the plume itself.

of the spectrum becomes weak above about 2000 K, this transition is suited for temperature measurement because the LIF is strong and the band spread is small. Figure 7.17(b) shows an example of a measured spectrum. By fitting a theoretical curve to this spectra, the rotational temperature of the YO molecules was found to be about 1000 K.

When a thin film is examined using an electron microscope, small particles with micrometre order size can be observed embedded in the film. Research on the generation mechanism of these particles is a subject of research because these particles degrade the film properties. Figure 7.18 shows the Mie scattering signal observed from such small particles obtained using a cw Ar-ion laser as the probe laser. When a particle moves through the probe beam, a spike appears in the signal due to Mie scattering from that particle. In conditions like this, when the spike corresponds to the signal from a single particle, the magnitude of the spike can be related to the size of the particle. As the measurement shown in figure 7.18 was done in vacuum conditions, it can be inferred that the small particles are directly ejected from the target surface. However, these particles move at much lower speeds than atomic and molecular species.

One-dimensional imaging measurements. It is possible to measure spatial distributions of particle densities by LIF techniques either by moving the position of the probe laser beam or by changing the position from which the LIF signal is detected. However, a preferable method is to measure the spatial distribution directly in a single laser shot using an array detector. In this experiment, a photodiode array (Hamamatsu C 2925) was used and a gated image intensifier (Hamamatsu PM 3323) was used to increase its sensitivity.

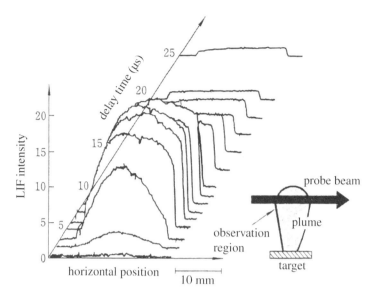

Figure 7.19. One-dimensional spatial distribution (in the horizontal direction) of LIF signals from YO molecules in the plume [16].

The signal was processed by a digital oscilloscope and then a microcomputer. In this way, not only single-shot measurements but also averaged profiles could be recorded. The image intensifier was gated with a 60 ns duration in order to reduce the effect of the background light emission.

Figure 7.19 shows the distribution of the LIF signal from YO molecules measured at a position 30 mm directly above the centre of the target, for vacuum conditions [16]. The TOF spectrum at each point and the time dependence of the LIF signal are shown. From the measured distributions, the angular dependence of the YO molecules ejected from the surface could be determined, and this was found to be extremely directional in the direction corresponding to $\cos^{13}\theta$.

In this kind of experiment, it is possible to expand the beam into a sheet-like beam, and measure the spatial distribution perpendicular to the target surface. Figure 7.20 shows the YO molecular density in the region 10–40 mm from the target surface when ablation was performed in an O_2 gas environment. Measurements are shown for two different pressures. For the relatively low-pressure 50 mTorr case, the particles move rapidly away from the target, diffusing as they move. For the higher pressure 300 mTorr case, the motion of the molecules released from the surface has virtually stopped by the time they are 30 mm from the surface, and the molecules simply diffuse after this time. In an actual deposition process, a substrate would be placed close to the target. The results shown in figure 7.20 indicate that, for high-pressure conditions, the particles ejected from the target surface would lose all of their initial energy through collisions with O_2 molecules in the background gas, and proceed to the substrate surface by diffusion alone.

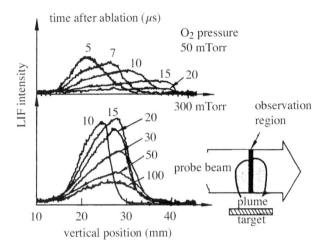

Figure 7.20. One-dimensional spatial distribution (in the vertical direction) of LIF signals from YO molecules in the plume [17].

Two-dimensional imaging measurements. In recent years, the development of CCD cameras as practical detection instruments and the corresponding development of image processing technology has made two-dimensional imaging possible. Adequate detection sensitivity for LIF measurements can be achieved by using an image intensifier in front of the CCD camera. By gating the image intensifier, images of very short-lived, essentially instantaneous phenomenon can be obtained. By recording these images in a computer, further image processing and data averaging are possible.

Figure 7.21 shows the results of a two-dimensional LIF measurement performed during PLD fabrication of YBCO films. In this case, a pulsed optical parametric oscillator was used as the probe laser [18]. LIF was observed at right angles to the sheet-like probe laser beam and the images were recorded by changing the delay time between the excimer laser and the probe dye laser. The experiment was performed in a 250 mTorr O_2 environment, and there was no substrate present. Although the experimental conditions are similar to those for the measurement shown in figure 7.20, much more information can be gained from these two-dimensional images.

The results show that Ba atoms have a relatively short decay time, and only exist in the region close to the target. This can be understood by considering the oxidation of Ba atoms in the O_2 environment, which leads to Ba atoms being converted to BaO molecules. A similar process occurs for Y atoms, and LIF images in figure 7.21(*b*) show the YO molecules on the surface of the expanding plume. In contrast to the Ba and Y cases, it can be seen that Cu atoms reach the substrate area without being oxidized. A gated CCD camera could be used to obtain, much more easily, images of emissive species but these LIF images are

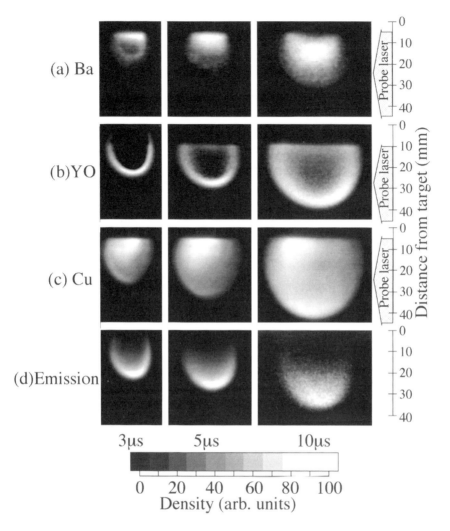

Figure 7.21. ICCD camera images of LIF from (*a*) Ba, (*b*) YO, (*c*) Cu and (*d*) emission from particles [18]. The ambient oxygen pressure was 100 mTorr.

much more valuable because these represent distributions of ground state species rather than excited species.

The method of rotational temperature measurement described previously and shown in figure 7.17 can be applied to make two-dimensional measurements. A temperature distribution, calculated from two images of LIF from YO molecules obtained using different excitation wavelengths, is shown in figure 7.22. It can be seen that heating at the shock front of the plume occurred initially, followed by cooling of the plume as it entered the diffusion phase.

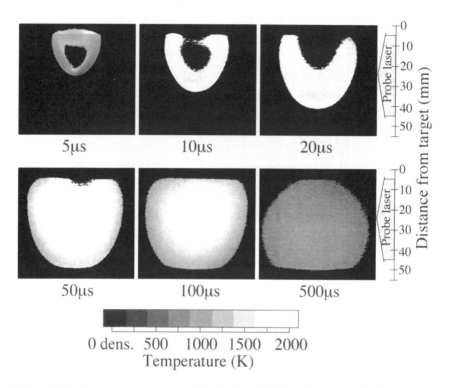

Figure 7.22. Rotational temperature distribution of YO molecules obtained from two different excitation wavelengths [13].

Another application of laser ablation is to generate nanometre size particles for photonics applications. Two-dimensional measurement techniques also have been used to study this process. Figure 7.23 shows images of Si nano-particles formed when a Si target was ablated in a He gas environment. The wavelength of the pulsed laser was tuned to a Si resonance at $\lambda = 250.69$ nm. The images show an umbrella-like cloud that forms after about 3 ms. When the laser wavelength was tuned away from the Si resonance wavelength, this cloud was still visible, indicating that the signal was due to Mie scattering from Si particles, rather from individual atoms. These results indicate that nano-particles were formed in this ablation process by condensation of Si clouds.

REFERENCES

[7.1] Ikeda M *et al* ed 1992 *Laser Process Gijutu Handbook* (Asakura Shoten) (in Japanese)
[7.2] Yardley J T 1985 *Laser Handbook* vol 5, ed M Bass and M L Stitch (North Holland) p 405
[7.3] Toyoda K and Murahara M 1986 *Excimer Laser Saisentan Ohyo Gijutu* (CMC) (in Japanese)

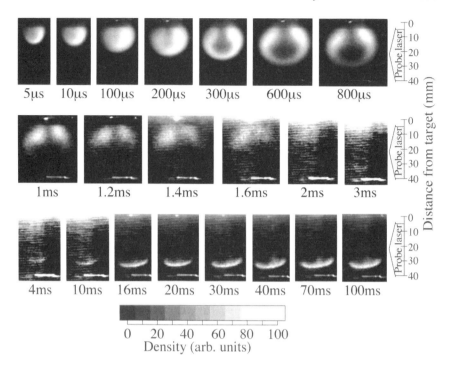

Figure 7.23. Si nano-particle formation process observed by a two-dimensional LIF method. The upper cloud was identified as being formed of Si atoms, and the lower umbrella-like cloud as Si nano-particles [19].

[7.4] Izawa Y 1987 *Denki Gakkaishi* **107** 1253 (in Japanese)
[7.5] Maeda M *et al* 1991 *Kogaku* **20** 2 (in Japanese)
[7.6] Reintjes J F 1985 *Laser Handbook* vol 5, ed M Bass and M L Stitch (North Holland) p 1
[7.7] Okada T *et al* 1989 *Laser Kenkyu* **17** 136 (in Japanese)
[7.8] Okada T *et al* 1987 *Appl. Phys.* **B44** 175
[7.9] Okada T *et al* 1988 *Appl. Phys.* **B47** 191
[7.10] Okada T *et al* 1990 *Appl. Phys. Lett.* **56** 1380
[7.11] Okada T *et al* 1992 *Japan. J. Appl. Phys.* **31** 3707
[7.12] Okada T and Maeda M 1993 *Oyo Butsuri* **62** 1221 (in Japanese)
[7.13] Okada T *et al* 1996 *Thermal Sci. & Eng.* **4** 27
[7.14] Okada T *et al* 1992 *Appl. Phys. Lett.* **60** 941
[7.15] Kumuduni W K A *et al* 1993 *Japan. J. Appl. Phys.* **32** L271
[7.16] Kumuduni W K A *et al* 1994 *Appl. Phys.* **B58** 289
[7.17] Okada T *et al* 1993 *Japan. J. Appl. Phys.* **32** L1535
[7.18] Nakata Y *et al* 1996 *J. Appl. Phys.* **80** 2458
[7.19] Muramoto J *et al* 1997 *Japan. J. Appl. Phys.* **36** L563

Chapter 8

Analytical Chemistry

8.1 Analytical Chemistry and Laser Spectroscopy

In recent years, instrumental analysis methods have become extremely widespread in the field of analytical chemistry. There are many types of instrumental analysis methods, but *spectroscopy-based techniques* are considered to be among the most powerful because they can provide both qualitative and quantitative detection of a large variety of atomic and molecular species. There are many advantages of using lasers as the light source for spectroscopic measurements, and these have been discussed often in previous chapters. From the viewpoint of analytical chemistry, the most important features of laser methods are the high detection sensitivity and the high spectral resolution that can be achieved [1]. This high sensitivity and high resolution are most easily achieved when the material being analysed is a gas, because atoms and molecules in a gaseous environment have narrow spectral features. When measurements of liquid and solid materials are desired, samples are evaporated so that spectroscopy measurement in the gas phase can be performed.

The spectral configuration of most molecules is extremely complicated, and the transitions in molecules often are observed in the form of spectrally broad 'bands' rather than spectrally narrow 'lines'. Hence, it can be difficult to take full advantage of the high spectral resolution achievable with laser methods. One technique that can be used to overcome this is to try to eliminate the Doppler-broadened component of the spectrum by injecting the sample into a high-velocity jet, and relying on adiabatic expansion of the jet to cool the species. To gain the best possible spectral resolution, however, the technique called *Doppler-free spectroscopy* is used. With this method, it is possible to achieve sub-Doppler width spectral resolution, so that spectral features such as hyper-fine structure and isotopic shifts can be resolved [2,3].

Many different laser spectroscopic methods have been applied in analytical chemistry, and the main techniques are listed in table 8.1. As can be seen from the table, most common laser spectroscopic techniques are used. Before the development of tunable wavelength lasers, methods such as Raman

Analytical Chemistry and Laser Spectroscopy

Table 8.1. Different types of laser spectral analysis methods used in analytic chemistry.

	Analysis method	Laser spectral analysis method
Methods using fixed-wavelength lasers	emission spectroscopy	laser microprobe method breakdown spectroscopy
	Raman spectroscopy	laser Raman spectroscopy
	mass spectroscopy	laser mass spectroscopy
Methods using tunable lasers	absorption spectroscopy	laser absorption spectroscopy intra-cavity absorption spectroscopy cavity ring-down spectroscopy (CRDS)
	fluorescence spectroscopy	laser-induced fluorescence (LIF) spectroscopy fluorescence lifetime measurement
	ionization spectroscopy	resonant ionization spectroscopy (RIS) optogalvanic spectroscopy (OGS)
	photo-thermal spectroscopy	photo-acoustic spectroscopy (PAS) thermal lens spectroscopy
	nonlinear spectroscopy	multiphoton ionization (MPI) spectroscopy coherent anti-Stokes Raman spectroscopy (CARS)

spectroscopy that used fixed-wavelength lasers were the most commonly used techniques, but methods based on tunable lasers have become more widespread in recent years. Despite their many advantages, however, the tunable laser based methods have not yet been developed into the kind of analysis instruments that are used widely in industry. This is due to the high cost and relatively complicated maintenance requirements of the lasers themselves. These problems, though, are gradually being overcome, and these methods are expected to become the most widely used analysis methods in the future, in industrial environments as well as in research fields.

Each of the methods listed in table 8.1 already has been described in previous chapters, and so, rather than discussing the methods themselves, the remainder of this chapter contains descriptions of practical measurements that illustrate the importance of laser spectroscopic methods in the field of analytical chemistry [1,3].

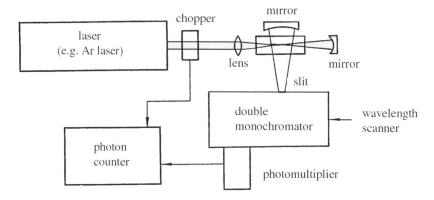

Figure 8.1 Schematic diagram of a laser Raman spectrometer.

8.2 Examples of Analysis using Laser Spectroscopic Techniques

8.2.1 Analysis using Laser Raman Spectroscopy

Raman scattering is a form of inelastic scattering of light from molecules. The scattered light is observed at a frequency that is shifted from the laser frequency v_l by an amount that depends on the energy of transitions in the target molecule. The scattered light frequency is $v_l \pm v_R$, and the frequency shift v_R is known as the Raman shift. The lower frequency, longer wavelength components, $v_l - v_R$, are called Stokes components and the higher frequency, shorter wavelength components, $v_l + v_R$, are called anti-Stokes components. Although Raman shifts due to electronic, vibrational and rotational transitions are all possible, Raman spectroscopy usually is based on Raman scattering from vibrational transitions.

The effectiveness of Raman scattering measurements has been improved greatly by the introduction of laser sources. Most commercially available Raman spectrometers are based on scattering from a fixed-wavelength laser. Figure 8.1 shows an example of such an instrument. An Ar-ion laser or a He–Ne laser usually is used as the light source. It is possible to obtain a Raman spectrum from all materials, solids and liquids as well as gases. In gases, the particle density is small and so multiple-reflection cells are used to direct the laser beam repeatedly through the gas sample. Even so, the Raman scattering signals are often extremely small, because Raman scattering cross-sections are extremely small, much smaller even than Rayleigh scattering cross-sections.

Several techniques can be used to improve the signal-to-noise ratio in Raman spectroscopy measurements. Most often, photon counting methods are used to detect the small signals, as is the case in the instrument shown in figure 8.1. By using a highly directional laser beam instead of a traditional lamp, the stray light can be reduced to an acceptable level. However, stray light is always the most serious problem in the measurement of Raman scattering. If a double

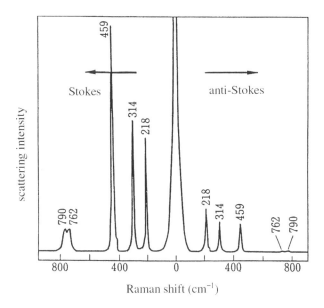

Figure 8.2. Raman spectrum of CCl_4 obtained using an Ar-ion laser ($\lambda=488$ nm) [4].

monochromator is used, as in the system shown in figure 8.1, the signals due to stray light, Rayleigh scattering and fluorescence at other wavelengths can be eliminated. By using an image-intensified diode array as the detector instead of a photomultiplier, a wide spectral region can be observed simultaneously. Also, use of a pulsed laser instead of a cw laser makes it possible to perform time-resolved Raman spectroscopy.

In general, Raman scattering cross-sections increase at shorter wavelengths. However, when short-wavelength lasers are used, the light tends to be absorbed by the gas, and dissociation of gas molecules might become a problem. To overcome this, an infrared Raman spectrometer combined with Fourier transform techniques has been developed, based on a 1.06 μm Nd:YAG laser. For the measurement of vibrational transitions of molecules, laser Raman spectroscopy can be considered to be complementary to infrared absorption spectroscopy.

Figure 8.2 shows an example of a Raman spectrum obtained when light from an Ar-ion laser ($\lambda = 488$ nm) was scattered from CCl_4 molecules [4]. The strongest central component, at the laser wavelength, is due to Rayleigh scattering and the peaks on either side are the Stokes and anti-Stokes components. The Stokes components at 218 cm^{-1}, 314 cm^{-1} and 419 cm^{-1} are due to fundamental vibrational modes of the CCl_4 molecule. This example is of Raman scattering from a gas, but scattering from solids also is possible. In that case, scattering from optical phonon, magnon and plasmon transitions can be observed.

262   *Analytical Chemistry*

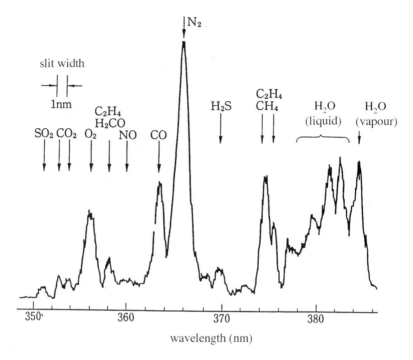

Figure 8.3. Raman spectrum obtained from smoke of burning oil, using a N_2 pulsed laser source ($\lambda=337$ mm) [5].

Raman spectroscopy is used for analysis of many different types of materials, and there are many examples of measurements of liquids and solids. One example of a gas measurement is the detection of Raman scattering from oil burning in atmosphere [5]. Figure 8.3 shows a spectrum measured in this study. A N_2 laser ($\lambda = 337$ nm) was used as the laser source, and a range of molecules, including SO_2, CO_2, CH_4, C_2H_4, H_2S, NO and CO, were detected.

8.2.2 Analysis using Laser-Induced Emission Spectroscopy

Historically speaking, spontaneous emission measurement techniques are the origin of all spectral analysis techniques. The traditional excitation sources used for these measurements include arc discharges, sparks, radio-frequency inductively coupled plasmas (ICP), and microwave-excited plasmas. Emission from excited atoms in the target species is measured, in a form of passive spectroscopy. By using lasers as excitation sources, the method becomes an active spectroscopy technique.

One example of a laser method is the so-called *laser microprobe technique*, in which a high-power pulsed laser, such as a Q-switched solid-state laser, is used to illuminate a material and cause a plasma to form on the surface. Emission from the laser-generated plasma then is detected and analysed. Figure

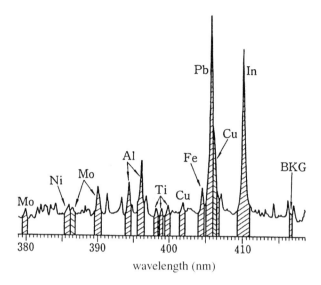

Figure 8.4. Emission analysis measurement of steel, obtained using the laser microprobe method [6].

8.4 shows an example of a spectrum measured using this method [6]. The spectrum was obtained when a steel surface was illuminated with light from a pulsed laser. More sensitive detection is possible using UV laser ablation techniques.

The remote sensing possibilities of this method were demonstrated in a different experiment, which monitored the effect of exposure to salt-containing winds on the insulators in high-voltage power transmission lines located near the ocean. The amount of salt (mostly NaCl) encrusted on the surface of insulators was measured, and the results were used to assess the damage of the insulating characteristics of the transmission line [7].

Another laser technique is *laser breakdown spectroscopy*, in which light from a powerful laser is focused to a small volume in a gaseous or liquid medium. The high light intensity causes a breakdown plasma to be produced, and emission from this laser-induced plasma is detected. Because breakdown usually is initiated by small particles in the medium, this method is well suited to *in situ* detection and analysis of small particles. Although high power density, of the order of 1 GW cm^{-2}, is required for breakdown, this can be readily produced by focusing the beam of a Q-switched Nd:YAG laser to a beam diameter of less than 1 mm. One example of such a measurement is the detection of elements such as Na, K, Ba, C and P in the atmosphere [8]. In this measurement, components of the buffer gas also were observed.

Although emission analysis has the advantage that different species can be detected simultaneously, in comparison with other spectroscopic techniques it is difficult to make any kind of quantitative measurement.

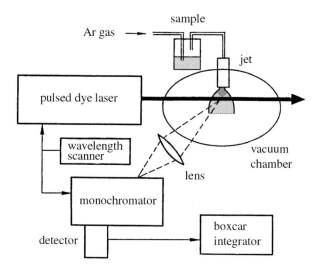

Figure 8.5. Schematic diagram of a synchronous scan LIF spectroscopic system combined with a molecular gas jet.

8.2.3 Analysis using Laser-Induced Fluorescence Spectroscopy

Absorption methods are the most important measurement techniques in the field of spectral analysis. These methods include infrared absorption, used for gas analysis, and atomic absorption, used for analysis of trace elements in liquids. By using a laser as the light source, very long optical path lengths become feasible, which improves the measurement sensitivity, and the light can be easily coupled to an optical fibre, which makes the method more convenient. However, there are several disadvantages. Absorption methods require very monochromatic light sources for achieving high sensitivity, and the operation and maintenance of tunable lasers is not as simple as that of traditional lamps. A particular problem is the lack of suitable tunable lasers for infrared wavelengths, which are necessary for detection of many molecules. Hence, the sensitivity of absorption methods is not necessarily improved by using a laser instead of a spectral lamp.

Laser-induced fluorescence (LIF) methods, in comparison, generally have far greater sensitivity than absorption techniques, for the reasons outlined in previous chapters. Two types of LIF methods are suitable for analysis purposes. In one method, the fluorescence light produced by the laser excitation is detected and analysed by a spectrometer. In the other method, the wavelength of the excitation laser is scanned, and the excitation spectrum of the LIF signal is measured.

In the first method, for the case of organic molecules that have spectrally wide absorption bands, fixed wavelength lasers at visible or UV wavelengths can be used to achieve high sensitivity. Lasers such as Ar-ion lasers, N_2 lasers

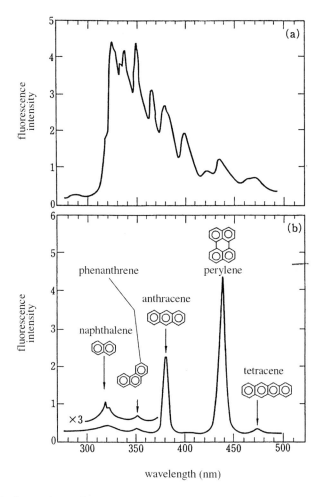

Figure 8.6. Comparison of spectra obtained from multiple-ring aromatic compounds using (*a*) a conventional LIF method and (*b*) a synchronous scan LIF method combined with a molecular jet [9].

and excimer lasers are suitable in this case. However, when the sample contains many species, it can be difficult to analyse the fluorescence spectrum because the spectral lines from different species overlap. In this case, the analysis can be simplified by using a gas jet to cool the sample and hence reduce the complexity of the spectrum. To simplify the spectrum even further, an apparatus such as that shown in figure 8.5 can be used, in which excitation and observation wavelengths are simultaneously scanned. This technique is called *synchronous scan luminescence*. Figure 8.6 shows a comparison of multiple-ring aromatic compound spectra, obtained using conventional LIF and synchronous scan luminescence [9].

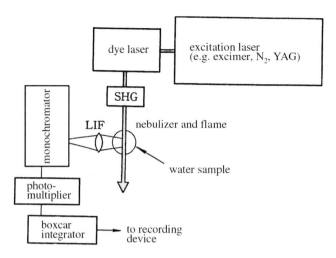

Figure 8.7. Schematic diagram of an atomic fluorescence analysis system using a flame, based on a tunable dye laser.

The most widely used analysis method is atomic absorption analysis using a flame, which can be used to make quantitative measurements of trace elements with high sensitivity, especially when the trace elements are in water. Atomic fluorescence analysis is regarded as a better technique in terms of sensitivity, but the lack of high-power monochromatic light sources has resulted in this method rarely being used. With the development of sources such as tunable dye lasers, many researchers are now attempting to develop analysis systems based on fluorescence measurements of flames [10,11].

An example of a flame-based atomic fluorescence system that uses a dye laser is shown in figure 8.7. A tunable wavelength range of 200–800 nm is desirable for the detection of a large variety of different elements, and so a dye laser pumped by a short-pulse laser is a suitable light source. The excitation laser could be an excimer laser, an N_2 laser or a Nd:YAG laser operated at the third-harmonic wavelength. A crystal for frequency doubling is necessary to extend the wavelength region into the ultraviolet. In the measuring system shown in figure 8.7, the sample is atomized using a burner with a nebulizer, in the same way as for the atomic absorption method. The laser is directed through the flame and the LIF signal is measured. Although the spectral width of the laser must be reasonably narrow, of the order of 1 cm^{-1}, the extremely narrow spectral width that is required for absorption measurements is not necessary.

In table 8.2, the detection limit for measurements of various trace elements are shown, in units of ng ml^{-1} (ppb), as compiled by Weeks *et al* [11]. The results shown in the column headed 'laser atomic fluorescence (A)' were obtained using a flame-based measurement system similar to that shown in figure 8.7. Excitation and observation wavelengths are shown on the right, but for some cases, these are the same, and so only one wavelength is shown. For most elements, the detection limit is about 1 ppb, which is a higher sensitivity than could be achieved with previous measurement techniques. As is done in atomic absorption analysis, the absolute concentration is determined by comparing the measured values with those obtained from a 'standard' sample of known concentration.

Various different methods exist for atomizing the samples to be measured. In flame-based atomic fluorescence analysis, the sample is usually atomized by using a burner with a nebulizer. In atomic absorption analysis, a graphite furnace is used to heat the sample to high temperature. In a different method, shown in table 8.2 as 'laser atomic fluorescence (B)', the sample is instantaneously heated using filaments, passed through a microwave discharge to atomize it, and then the LIF measurement is performed [12]. For many elements, very high sensitivity has been achieved with this method. In addition to these techniques, atomization using a graphite cup [13] and by laser ablation [14] also are being investigated.

Table 8.2. Detection limits of trace element in water for measurement by atomic absorption analysis [11]. The units are ng ml^{-1}.

Element	Atomic absorption (flame)	Atomic absorption (furnace)	Conventional atomic fluorescence	Laser atomic fluorescence (A)	Laser atomic fluorescence (B)	Wavelength (nm)
Ag	1	0.01	0.1	4	—	328.1
Al	30	0.1	100	0.6	—	394.4/396.1
Ba	20	0.6	—	8	18.6	553.7
Bi	50	0.4	5	3	8.4	306.8
Ca	1	0.04	20	0.08	1.95	422.7
Cd	1	0.008	0.001	8	—	228.8
Co	2	0.2	5	1000	3.4	230.9
Cr	2	0.2	5	1	—	359.3
Cu	1	0.06	0.5	1	0.6	324.7
Fe	4	1	8	30	4.35	296.7/372.5
Ga	50	0.1	10	0.9	—	403.3/417.2
In	30	0.04	100	0.2	0.84	410.4/451.1
Li	1	0.3	—	0.5	0.8	670.8
Mg	0.1	0.004	0.1	0.2	0.21	285.2
Mn	0.8	0.02	1	0.4	45.4	279.5
Mo	30	0.3	500	12	—	379.8
Na	0.8	—	—	<0.1	0.027	589.0
Ni	5	0.9	3	2	3.6	361.0/352.4
Pb	10	0.2	10	13	1.14	283.3/405.8
Sr	5	0.1	30	0.3	—	460.7

(A) atomic fluorescence spectroscopy using a flame (Weeks S J, Hamaguchi H and Winefordner J D 1978 *Anal. Chem.* **50** 360).

(B) atomic fluorescence spectroscopy using a microwave discharge (Oki Y *et al* 1993 *Anal. Chem.* **65** 2096).

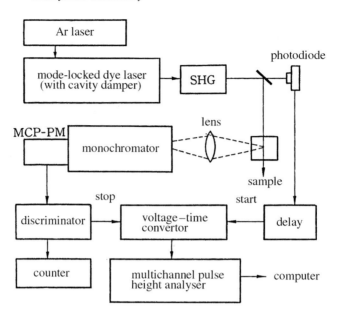

Figure 8.8. Schematic diagram of an LIF measuring apparatus with time resolution of ps order [15].

Another important area in which LIF methods are used is the study of the dynamic structure of molecules. Pulsed excitation is used for measurements of fluorescence lifetime or the time dependence of the fluorescence spectrum. In recent years, this new research field is progressing rapidly, largely due to the development of lasers with femtosecond pulse lengths.

Figure 8.8 shows an example of an LIF detection system capable of picosecond order temporal resolution [15]. A mode-locked dye laser with pulse width of a few picoseconds was used as the light source. If only nanosecond order resolution was required, a compact inexpensive N_2 laser could be used as the laser source. The detector was a multichannel-plate (MCP) photomultiplier with a fast time response. A time-correlated photon-counting method triggered in phase with the excitation and fluorescence light pulses was used. The time response of this kind of electronic device is limited, and so temporal resolution of less than about 100 ps was difficult to achieve. In order to observe faster phenomena, a nonlinear correlation method or a streak camera detector must be used.

8.2.4 Analysis using Laser Ionization Spectroscopy

There are two very different spectroscopic methods that use lasers as the ionization source. There is a non-resonant method, shown in figure 8.9(*a*), and a resonant method, shown in figure 8.9(*b*). The non-resonant method cannot be used to distinguish between different species, and so usually is used together

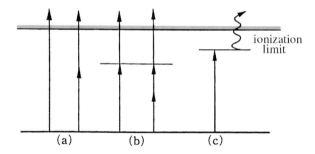

Figure 8.9. Laser ionization method. (*a*) shows a non-resonant version, (*b*) shows a resonant version and (*c*) shows the optogalvanic effect.

with mass spectroscopy. Hence, the non-resonant method can be viewed as a type of mass spectroscopy technique.

Resonant ionization spectroscopy (RIS) uses a tunable laser to excite the atom to an intermediate energy level, as shown in the figure. The resonant step provides species selectivity. One example that shows the detection limits achievable with this technique is the experiment of Hurst *et al*, who used a two-step ionization method together with detection by a proportional counter to measure Cs atom density [16]. Measurement down to a single Cs atom was possible. Because ionization cross-sections for atoms are much smaller than absorption cross-sections, pulsed lasers with high instantaneous powers are required for this type of measurement. If the instantaneous power is extremely high, of the order of 10^9 W cm^{-2}, nonlinear multiphoton ionization can occur, as shown in figure 8.9.

RIS has several advantages for analysis measurements. These include high sensitivity, no dependence on plasma emission because the signal is obtained as an electrical signal, and the capability to detect particles that do not fluoresce. In vacuum conditions, a highly sensitive electron multiplier, called a channeltron, can be used to detect ions. When there is a background gas present, the simplest detection method is just to insert an electrode into the gas and measure the ion current.

Figure 8.10 shows an example of an RIS measurement apparatus, used for measurements of organic molecules evaporated in a buffer gas [17]. A pair of planar electrodes, separated by 6 mm, were positioned inside the ion cell, and the ionization current produced by a pulsed laser was detected and then analysed by a boxcar averager. For benzene derivatives, the detection limit was at the ppb level. For other materials, ppt detection is possible by using a furnace to atomize the material, and then using an electron beam to ionize the atoms in a vacuum.

A related method is called *optogalvanic spectroscopy* (OGS). In this method, shown in figure 8.9(*c*), an atom or molecule is excited to an intermediate level by a laser, and then ionized by a thermal or discharge effect. Thus, OGS can be considered to be a type of laser ionization spectroscopy. In the most widely known application of OGS, a laser excites atoms in a small

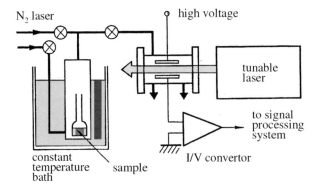

Figure 8.10. Schematic diagram of an RIS apparatus for measurement of organic molecules in the gas phase.

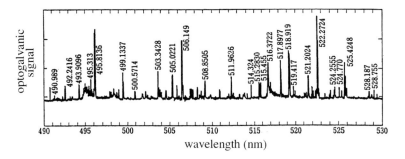

Figure 8.11. Example of an OGS spectrum observed from a hollow cathode discharge tube.

discharge tube, and the change in discharge impedance due to the subsequent ionization is measured [18]. This arrangement is used as a spectral standard because the wavelengths of observable spectral lines range from the ultraviolet to the near-infrared. Transitions of the particles of the discharge gas (usually a noble gas) and from atoms of the electrode material can be detected, and excitation from both ground state and metastable species is possible. Figure 8.11 shows an example of a spectrum obtained from a hollow cathode discharge: all the identified spectral lines correspond to transitions of the argon buffer gas.

For analysis purposes, OGS can be used in several arrangements. It is technically possible to introduce the sample to a discharge and then observe the optogalvanic excitation spectrum. The most practical method, however, is to measure the OGS spectrum for species in a flame, rather than in a discharge. An example of this type of measurement apparatus is shown in figure 8.12 [19]. The flame is used to atomize the sample, and small electrodes are placed near the flame. Atoms that have been excited to intermediate levels by the flame are

Examples of Analysis using Laser Spectroscopic Techniques 271

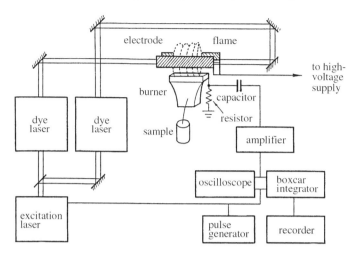

Figure 8.12. Schematic diagram of an LEI spectral analysis system that can be used for two-step excitation [19].

selectively ionized by tuning the wavelength of the laser, and the current between the electrodes is measured. This commonly is called *laser-enhanced ionization* (LEI) *spectroscopy*, but can be considered to be one form of optogalvanic spectroscopy. The system shown in figure 8.12 has two dye lasers in order to use a two-step excitation method, but in its more conventional form, only one dye laser is used.

Table 8.3 shows the lower detection limits in ng ml^{-1} (ppb) for measurements of trace elements in water by the LEI system shown in figure 8.12 [19]. Not only was the limit for most species of the 1 ppb order, but for some species (Li, In, Na, Tl) ppt order was achieved. In most cases, this high sensitivity was achieved by using two-step excitation to excite the atoms to high energy levels. Introducing an additional beam from a YAG laser can increase the photo-ionization signal further, and increases in signal of a factor of more than 100 have been reported [20].

8.2.5 Analysis using Laser Photothermal Spectroscopy

In *photothermal spectroscopy*, an object is illuminated with light and absorption of that light causes local heating of the material. If the heating is continued, there will be an optical, thermal or elastic change in the material, and this change can be detected [3]. Although it is possible to use light sources other than lasers, laser sources are particularly well suited for this technique because a high intensity laser can provide very localized heating. Photothermal techniques have become very effective since the development of suitable laser sources.

Two types of photothermal spectroscopy are widely used. These are photo-acoustic spectroscopy, in which the sound generated by the local heating is detected, and thermal lens spectroscopy, in which the change in refractive index induced by the local heating is detected. Both methods have the advantages that it is possible to measure gaseous, liquid and solid samples, and no special pretreatment of the sample is necessary. In addition, these methods have significantly higher sensitivity than ordinary absorption techniques.

Photo-acoustic spectroscopy. If intermittent light is used to illuminate an object, absorption of the light will change the pressure in that region, and an acoustic wave will be generated at the same frequency as the intermittent light. This phenomenon is called the photo-acoustic effect, and has been known since the 19th century. *Photo-acoustic spectroscopy* (PAS), based on this principle, has a fairly long history as a gas analysis technique, but it was the work reported by Kreuzer in 1971, showing that ppb order sensitivity was possible, that really demonstrated the effectiveness of this technique [21].

An example of a PAS arrangement based on a cw laser is shown in figure 8.13. An acoustic wave is generated inside the cell, and the chopper and microphone are positioned to optimize the size of the signal. Acoustic waves also are generated as the laser passes through the windows of the cell, and these waves are sources of noise. The detected acoustic signal is analysed using a lock-in amplifier.

A pulsed laser is used as the excitation source for cases when a laser that is tunable over a wide wavelength range is necessary. Although the laser pulse generates a complicated acoustic wave as it enters the cell, the signal can be averaged using a boxcar averager with the averaging window optimized in order

Table 8.3. Detection limits of trace elements in water for measurement by LEI spectroscopy using a flame [19]. The units are ng ml^{-1}.

Element	One-photon excitation	Two-photon excitation	Element	One-photon excitation	Two-photon excitation
Ag	1	0.05	K	0.1	—
Al	0.2	—	Li	0.001	0.0002
As	3000	—	Mg	0.003	0.001
Au	1.2	1	Mn	0.04	0.02
Ba	0.2	—	Mo	10	—
Bi	0.2	—	Na	0.001	0.0006
Ca	0.006	0.03	Ni	0.02	0.04
Cd	0.2	0.1	Pb	0.2	0.09
Co	0.06	0.08	Rb	0.09	—
Cr	0.2	0.3	Sn	0.4	0.3
Cs	0.002	—	Sr	0.003	0.2
Cu	3	0.07	Ti	1	—
Fe	0.08	0.1	Tl	0.006	0.008
Ga	0.01	—	Yb	1.7	0.1
In	0.001	0.0009	Zn	—	1

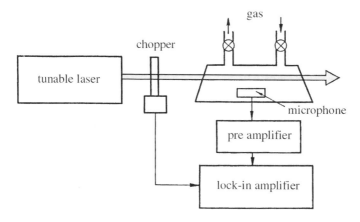

Figure 8.13. Schematic diagram of a PAS measurement system based on a cw laser.

to give the best signal-to-noise ratio. In order to improve the signal-to-noise ratio, attachment of a resonance tube on the microphone is often used effectively.

Table 8.4 shows the detection limits for measurements of trace gases that are present in the atmosphere [1]. Except for SO_2 and NO_2, these require infrared laser sources. This can be a potential problem because the sensitivity of the PAS technique increases greatly as the laser power is increased, but until recently there have been few high-power infrared sources. Recently, optical parametric oscillators (OPO), which are tunable over a wide wavelength range in the infrared and are capable of high output power, have been developed as practical laser sources. With these new laser sources, PAS methods are expected to become important analysis techniques for atmospheric gases.

Thermal lens spectroscopy. Another important photo-thermal spectroscopic

Table 8.4. Detection limits of trace molecules in the atmosphere for measurement by PAS techniques.

Molecule	Detection limit (ppb)	Excitation laser
SO_2	0.1	dye laser (UV)
CO	150	CO Laser
NO_2	10	cw dye laser
	0.1	CO laser
NO	0.4	CO laser
H_2O	14	CO_2 laser
CH_4	10	He–Ne laser (IR)
C_6H_6	3	CO_2 laser

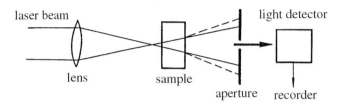

Figure 8.14. Basic arrangement for thermal lens spectroscopy.

technique is *thermal lens spectroscopy* [3,22]. The basic arrangement for this method is shown in figure 8.14. A laser beam with a Gaussian beam shape is focused into a sample, causing the sample to be locally heated by absorption of the laser light. The refractive index of the sample is changed by the temperature rise, and so the beam diverges from the heated sample, which acts as a kind of concave lens.

Several different variations of this arrangement exist. One of these is the thermal light-deflection method, in which a separate probe laser is used to measure the beam deflection. In another method, called the thermal light-diffraction method, a thermal diffraction grating is formed by interference in the sample. Both of these techniques have been developed into measurement instruments [3]. The arrangement shown in figure 8.14, however, has several advantages. One is that the sensitivity can be increased by using a separate probe laser on the same beam path as the intense laser. Another is that a long cell can be used, which also increases the sensitivity. In addition, it is claimed that even higher signal-to-noise ratios can be obtained by using a CCD camera to measure the beam cross-section, instead of only measuring the central intensity, as shown in figure 8.14.

The sensitivity of thermal lens spectroscopy depends on the temperature coefficient of the refractive index, the thermal conductivity of the material, and also on the laser power. Although both cw and pulsed lasers can be used in this method, it has been shown theoretically that the highest sensitivity can be obtained by using a cw laser when the sample is liquid, and a pulsed laser when the sample is a gas. Although there are relatively few examples of analysis of atmospheric gases, a detection limit of 5 ppb has been reported for NO_2, using a system with an Ar-ion laser and 1 m length cell [23].

REFERENCES

[8.1] Klinger D S ed 1083 *Ultrasensitive Laser Spectroscopy* (Academic Press)
[8.2] Demtröder W 1981 *Laser Spectroscopy* (Springer-Verlag)
[8.3] Imasaka T and Ishibashi N 1990 *Progress in Quantum Electronics* **14** 131
[8.4] Hamaguchi H *et al* ed 1992 *Fundamentals and Applications of Laser Spectroscopy* (IPC Ltd) p 332 (in Japanese)
[8.5] Inaba H and Kobayashi T 1972 *Opto-Electronics* **4** 101

- [8.6] Cremers D A 1987 *Appl. Spectroscopy* **41** 572
- [8.7] Fujiyoshi S *et al* 1992 *Laser Kenkyu* **20** 955 (in Japanese)
- [8.8] Loree T R and Radziemski L J 1981 *Plasma Chemistry and Plasma Processing* **1** 271
- [8.9] Imasaka T *et al* 1998 *Anal. Chem.* 1346
- [8.10] Fraser L M and Winefordner J D *Anal. Chem.* **44** 1444
- [8.11] Weeks S J *et al* 1978 *Anal. Chem.* **50** 360
- [8.12] Oki Y *et al* 1933 *Anal. Chem.* **65** 2096
- [8.13] Falk H *et al* 1988 *Spectrochemica Acta* **43B** 1101
- [8.14] Oki Y *et al* 1994 *Opt. Comm.* **110** 298
- [8.15] Hamaguchi H *et al* ed 1992 *Fundamentals and Applications of Laser Spectroscopy* (IPC Ltd) p 500 (in Japanese)
- [8.16] Hulst G S, Nayfeh M H and Young J P 1977 *Phys. Rev.* **A15** 2283
- [8.17] Yamada S *et al* 1988 *Bunseki Kagaku* **37** 216 (in Japanese)
- [8.18] King D S and Schenck P K 1978 *Laser Focus* March p 50
- [8.19] Axner O *et al* 1987 *Appl. Opt.* **26** 3521
- [8.20] Oki Y *et al* 1994 *Opt. Comm.* **110** 105
- [8.21] Kreuzer L B 1971 *J. Appl. Phys.* **42** 2934
- [8.22] Imasaka T and Ishibashi N 1983 *Laser Kenkyu* **11** 129 (in Japanese)
- [8.23] Higashi H, Imasaka T and Ishibashi N 1983 *Anal. Chem.* **55** 1907

Chapter 9

Remote Sensing

9.1 LIDAR and Monitoring of the Atmosphere

Optical methods have an important role in remote sensing technology. These techniques can be divided into active and passive methods, depending on whether or not the person making the measurement first transmits a signal as part of the measurement process. To a certain extent, any activity in which a telescope or a camera or even just the human eye is used to view some object at a distance can be considered to be a form of passive remote sensing. If a variety of physical parameters can be determined by measuring spectral information about radiation coming from the object. It is possible to obtain global information of this kind by using aeroplanes or satellites.

Compared with passive remote sensing, laser methods are active techniques in which the transmitted light scatters or reflects from something in the atmosphere, and the reflected or scattered light is analysed to provide information about some physical quantity. Of the many quantities that can be measured, physical distance is perhaps the most important, and various kinds of laser ranging techniques have been developed. *Laser radar* is one such technology, in which a pulse of light is transmitted, and temporal analysis of the scattered light intensity provides information about the shape and/or distribution of objects in the atmosphere. Conventional radar (*ra*dio *d*etection *a*nd *r*anging), which usually uses microwaves, is one of the most important remote sensing techniques. Laser radar, which uses light instead of microwaves as the radiation, also is known as LIDAR (*l*ight *d*etection *a*nd *r*anging) [1,2].

There are several important differences between microwave-based methods and laser-based methods. In general, microwaves pass straight though clouds and smog whereas laser light in the visible or near-visible wavelength ranges scatters greatly. This means that LIDAR cannot be used to detect any matter which is hidden by clouds or smog, but conversely, it is possible to measure the spatial distributions of the small particles which make up the clouds and smog. By using short-wavelength lasers, LIDAR can even be used to detect the presence of molecules in the atmosphere by Rayleigh scattering. This ability to measure spatial distributions of specific atoms and molecules is one of the key

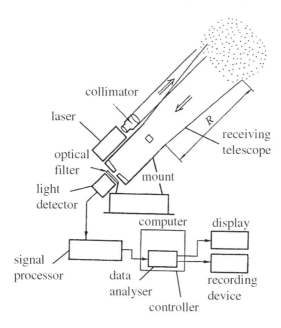

Figure 9.1. Basic arrangement for a LIDAR measurement system.

advantages of remote sensing using LIDAR methods.

LIDAR research began in the year following the invention of the ruby laser in 1961 and so has a long history. One reason for this is that LIDAR, together with similar microwave techniques, is suitable for monitoring the weather and the general atmospheric environment. LIDAR techniques are able to measure the spatial distributions of particles high in the atmosphere, at altitudes of up to 100 km, and this makes them very powerful and attractive techniques for monitoring the atmosphere, especially in recent times when global environmental problems have become very important.

9.1.1 LIDAR Theory

Figure 9.1 is a schematic diagram of a generic LIDAR measurement system. A laser beam with pulse length τ is collimated and sent into the atmosphere, the backscattered light is received by a large telescope, and the signal intensity is time-analysed and recorded. As the speed of light c is known, the scattered power $P(R)$ returned from a point at distance R can be determined by the following equation [1,2]:

$$P(R) = P_0 K\left(\frac{c\tau}{2}\right) A\beta(R)Y(R)T_0(R)^2 / R^2. \qquad (9.1)$$

In this expression, P_0 is the power of the transmitted beam, K is the efficiency of the receiving system, A is the effective area of the telescope entrance, $\beta(R)$ is the differential cross-section for backward scattering and $Y(R)$ is a geometrical factor determined by the overlap of the transmitted beam and the receiving telescope. If α is the distinction coefficient, σ the absorption cross-section and N the density of the absorbing molecules, then the one-way transmission efficiency $T_0(R)$ can be expressed as

$$T_0(R) = \exp\left[-\int_0^R \{\alpha(R') + \sigma N(R')\} dR'\right]. \qquad (9.2)$$

If the response of the detection system is sufficiently fast, the resolution in distance can be as high as $c\tau/2$. The amount of light that is scattered from the atmosphere usually is very small, however, and in many cases, the detected signal is integrated over a certain period of time in order to improve the signal-to-noise ratio. In these cases, the time response of the detection system is deliberately slowed down and the resolution in distance is determined by the integration time. Also, again to improve the signal-to-noise ratio, it is common for the signal to be integrated or averaged for a large number of transmitted laser pulses.

The type of detector used to receive the signal depends on the wavelength range of the measurement. Sensitive photomultiplier tube detectors can be used when the wavelength is less than 1 μm. Photon-counting methods are often used in this wavelength region. For the wavelength range of 1–2 μm, avalanche photodiodes can be used, and, in general, semiconductor light detectors can be used for wavelengths longer than 1 μm. In these methods, the signal waveform is integrated for each pulse, with the zero in time being determined by the time of transmission. The signal-to-noise ratio of the detected signal is determined by the total number of transmitted photons.

9.1.2 Different LIDAR Techniques

Table 9.1 is a list of different types of LIDAR techniques. Because Mie scattering and Rayleigh scattering occur simultaneously, it is difficult to separate the contributions from each other, and some additional information is necessary in order to analyse the data properly. Using equation (9.1) and a wavelength at which there is no absorption by atmospheric molecules, $\beta(R)$ or $\alpha(R)$ can be

Table 9.1. Different LIDAR methods and the species measured by each.

Type of system	Measured species
Rayleigh scattering LIDAR	atmospheric molecules (N_2, O_2)
Mie scattering LIDAR	aerosols, clouds, smog, mist
DIAL	NO_2, SO_2, O_3, H_2O, HCl, Cl_2
Fluorescence LIDAR	Na, K, Na, Ca, Ca^+, NO_2, OH
Raman LIDAR	N_2, CO_2, H_2O, SO_2

determined from the detected signal $P(R)$. From this, the distribution of scattering particles can be found. For this analysis, we must assume some kind of known relationship between α and β. The intensity of Rayleigh scattering varies inversely with the fourth power of the wavelength, and so the Rayleigh scattering intensity is very large at short wavelengths. The wavelength dependence of Mie scattering depends on the size distribution of the scattering particles.

LIDAR techniques for measuring the spatial distribution of a specific atomic or molecular species are based on three spectroscopic techniques: absorption, fluorescence and Raman scattering. Amongst these, the absorption method known as DIAL (differential absorption LIDAR) is the most widely used.

In the DIAL method, absorption at two different wavelengths is detected. Two laser beams, one with a wavelength corresponding to a strongly absorbing transition, λ_1, and one with a wavelength at which there is only weak absorption, λ_2, are transmitted into the atmosphere. These beams are scattered by particles in the atmosphere, due to Rayeigh or Mie scattering, and the respective backscattered signals, $P_1(R)$ and $P_2(R)$, are measured. The density distribution of the absorbing species can be determined from the difference between the two signals. This density distribution can be determined using a calculation procedure based on equation (9.1). The average particle density, $N(R)$, in the distance region between R and $R + \Delta R$, is given by the following expression:

$$N(R) = \frac{1}{2(\sigma_1 - \sigma_2)\Delta R} \left\{ \ln \frac{P_1(R)}{P_1(R + \Delta R)} - \ln \frac{P_2(R)}{P_2(R + \Delta R)} \right\}. \quad (9.3)$$

In this equation, σ_1 and σ_2 are the absorption cross-sections for λ_1 and λ_2 respectively. A variety of species have been measured in this way. O_3 and SO_2 have been detected using wavelengths in the ultraviolet part of the spectrum. In addition, NO_2 (visible wavelengths), H_2O, CH_4 and HCl (infrared wavelengths) have been detected.

In fluorescence methods, quenching effects due to particle collisions are usually quite severe, and there are relatively few applications suitable for these techniques. NO_2 has been detected using wavelengths in the visible part of the spectrum, and the OH radical, using near-ultraviolet wavelengths. However, it is in the detection of the thin layers of metal atoms, such as Na, K, Li, Ca and Ca^+, that exist at altitudes of above 100 km, where quenching is negligible, that the strength of this method has been best demonstrated. The reason for this is that the cross-sections for fluorescence are so large that extremely low particle densities, down to the order of 100–1000 particles per cm^3, can be detected. In addition to these applications, remote sensing utilizing the fluorescence from water surfaces, and from chlorophyll, have been attempted.

In Raman LIDAR, the beam from a fixed-wavelength, solid-state laser is sent into the atmosphere and the backscattered light is spectrally analysed. It is theoretically possible to detect a wide variety of molecules with this technique, but Raman scattering cross-sections are so small that there are relatively few species in the atmosphere that can be detected. It is really only possible to

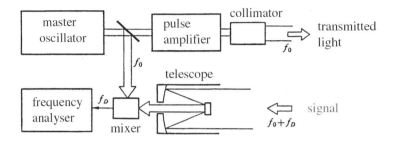

Figure 9.2. Schematic diagram of a Doppler LIDAR system.

measure species that have relatively high densities, such as N_2, O_2, CO_2 and H_2O.

In all of the above methods, the intensity of the backscattered signal is the only important quantity. In Doppler LIDAR, however, the Doppler shift of the scattered light is measured and from this, the wind speed can be determined. Figure 9.2 shows a schematic diagram of a Doppler LIDAR system. For this type of measurement, a laser with a high degree of coherence and a heterodyne detection system are required. Heterodyne detection is an effective method of increasing the signal-to-noise ratio, and is especially attractive for measurements using infrared wavelengths, for which high-quality detectors are not available [3].

Although LIDAR can be considered a powerful method for remote sensing of the atmosphere, one problem is that the distance that can be measured is severely restricted during daytime hours and also when the weather is poor. In addition, it is only of limited usefulness to analyse data that is obtained at a single location. With regard to these problems, it is possible to conceive compact LIDAR systems that can be mounted in trucks or aeroplanes, or a network of measuring stations that would allow data to be obtained over a wide area. For this to be feasible, the development of a compact LIDAR system with a high degree of reliability, probably based on an all solid-state laser, is necessary.

The ideal LIDAR system would be one that was mounted in a man-made satellite and directed inwards towards the Earth's surface. Not only would this system be unaffected by weather conditions, it also would be able to obtain global data in a short time. The first such space-LIDAR experiment was successfully conducted using the space shuttle in September 1994 [4]. With the increasing importance of global atmospheric and climatic research, further development of space-LIDAR systems can be expected.

9.2 Representative LIDAR Experiments

9.2.1 Mie Scattering LIDAR

Mie scattering LIDAR is used typically for the measurement of aerosol distributions, mixtures and inversion of layers in the atmosphere, transport and dispersion of factory smoke, and, using aerosols as a tracer, measurement of wind speed. It also is used for measurements of low altitude clouds and general visibility levels.

Flashlamp-pumped, Q-switched Nd:YAG lasers are the optimal laser sources for Mie scattering LIDAR. The second-harmonic wavelength of 532 nm, rather than the fundamental wavelength of 1.06 μm, is more suitable because of the low noise photomultiplier detectors that are available for visible wavelengths. The specifications of such an instrument, the LIDAR system at the National Institute of Environmental Studies (NIES) in Japan, are given in table 9.2. Measurements of the distribution of aerosols (μm size particles in the atmosphere) within a 50 km field of view have been made with this system [5].

Figure 9.3 shows the aerosol distribution in the Lake Kasumigaura region near Tokyo, measured using this LIDAR system [5]. The measurement was made by directing the laser beam in the horizontal direction and then sweeping the beam horizontally. In this graph, the quantity called aerosol concentration is the backscattered light measured for each azimuthal angle, corrected for the reduction in intensity due to the length of the optical path. The signal from 33 750 laser pulses was measured, and the measurement time was 22.5 min. The total region measured had an area of about 20×20 km^2, and the data were obtained by sweeping the beam through 90 degrees. In the figure, the size of one pixel is 100×100 m^2. The regions that show high aerosol concentrations are downwind of factories and urban areas, and this kind of data can be used to understand the source of aerosols and the direction of the wind.

By directing the laser beam vertically, the altitude distribution of aerosols over the course of one day was measured, and this is shown in figure 9.4 [6]. The data have been plotted using 'edge enhancement' so that sharp changes in the aerosol distribution can be easily seen. The data show that atmospheric

Table 9.2. Equipment specifications of the wide-range Mie scattering LIDAR at the National Institute of Environmental Study (NIES), Japan [5].

Item	Specification
Laser	Nd:YAG laser
wavelength	1.06 μm / 532 nm
output energy	1.2 J/pulse (1.06 μm)
	0.4 J/pulse (532 nm)
pulse width	15 ns
beam divergence	0.3 mrad
maximum repetition rate	25 Hz
Detection telescope	Cassegrain-style reflector telescope
effective aperture	1.5 m diameter
scanning ability	vertical and horizontal
scanning precision	0.3 mrad

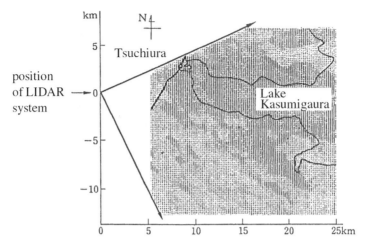

Figure 9.3. Aerosol distribution obtained by horizontal scattering in the NIES Mie scattering LIDAR system [5].

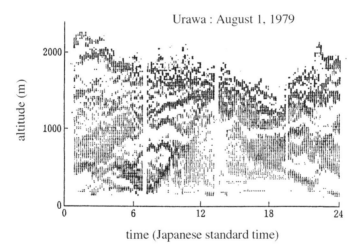

Figure 9.4. Altitude distribution of aerosols in the atmosphere measured over a 24 h period by a Mie scattering LIDAR system [6].

pollutants that originate at the Earth's surface are well mixed in a kind of atmospheric boundary layer, with the density then decreasing sharply at altitudes higher than this layer. This boundary structure can be seen clearly in this experimental result, which is a good example of the useful type of measurement that can be performed using aerosols as a tracer.

Another interesting measurement made using Mie scattering LIDAR concerns aerosols in the stratosphere, at altitudes above 10 km. These aerosols originate mainly from the sulphides that are released during volcanic eruptions.

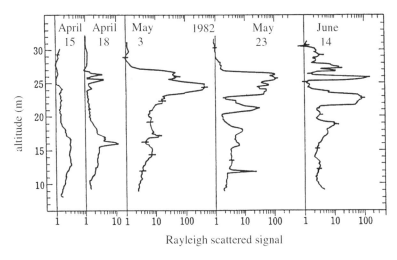

Figure 9.5. Measurements made using a Mie scattering LIDAR system at Fukuoka, Japan, showing the increase in aerosols in the stratosphere two weeks after the eruption of the El Chichon volcano [7].

The aerosol density in the stratosphere increased sharply after the Mexican volcano, El Chichon, erupted in April 1982, and then also after the eruption of Mt Pinatubo in the Philippines in June 1991. It was observed that the amount of aerosols in the stratosphere increased globally for several months after the eruptions, and then gradually reduced over a period of 3 to 4 years. There also was some indication that the blocking of the solar radiation, known as the 'parasol effect' that occurs during this time, produced abnormal weather patterns.

The dramatic increase in aerosol concentrations that occurred on these two occasions was recorded in detail by LIDAR systems around the world, and valuable data regarding atmospheric transport around the Earth were obtained. Figure 9.5 shows the first measurement of this aerosol increase, obtained two weeks after the El Chichon eruption by the Nd:YAG based LIDAR instrument at Kyushu University in Fukuoka, Japan [7].

9.2.2 Rayleigh Scattering LIDAR

Rayleigh scattering LIDAR is used principally to determine the variation of atmospheric pressure with altitude. This is achieved by measuring the intensity of the backscattered Rayleigh signal. For altitudes less than about 30 km, the presence of large amounts of aerosols results in Mie scattering, and it is difficult to separate the Rayleigh and Mie scattered components in the detected signal. For higher altitudes, however, up to 80–100 km, accurate information about the atmospheric pressure can be obtained from the Rayleigh scattered signal. Rayleigh scattering cross-sections become larger at short wavelengths but if the

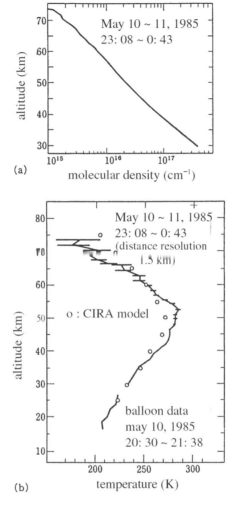

Figure 9.6. Example of atmosphere measurements made by Rayleigh scattering LIDAR, using a XeF laser. (*a*) shows the molecular density distribution and (*b*) shows the temperature distribution calculated from the results in (*a*) [9].

wavelength is too short, absorption by ozone molecules in the atmosphere restricts the measurable distance. The Nd:YAG laser, operated at its second-harmonic wavelength (532 nm), and the XeF excimer laser (351 nm) are the main laser sources used for these measurements [8,9].

Figure 9.6(*a*) shows an altitude distribution of atmospheric pressure measured using a XeF Rayleigh LIDAR [9]. The temporal change of this profile was compared with a standard atmospheric model (CIRA) and analysed in terms of atmospheric transport. This analysis showed that significant fluctuations in the atmosphere had occurred. Furthermore, the atmospheric temperature profile

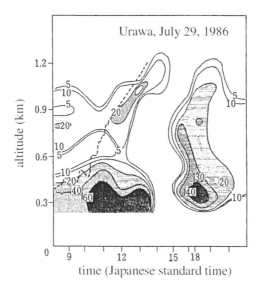

Figure 9.7. Example of an altitude distribution of NO_2 molecules in the atmosphere measured using DIAL (the units are ppb) [5].

also could be determined from the altitude distribution of the density profile, using the ideal gas equation and the hydrostatic pressure equation [9]. Figure 9.6(b) shows the temperature profile calculated in this way. The data below 30 km were obtained using a balloon for meteorogical sounding. Above this altitude, where balloon observation cannot be used, the data obtained by LIDAR are very useful.

9.2.3 Differential Absorption LIDAR (DIAL)

Amongst the spectroscopic methods that are used to measure spatial distributions of specific molecules, the DIAL method is the most widely used technique. Molecules such as NO_2, SO_2 and O_3 now are measured routinely using absorption transitions in the visible and near-ultraviolet. In the near-infrared, where there is a lack of powerful tunable laser sources, only H_2O can be measured in a similarly routine manner. Through the development of optical parametric oscillators (OPO), however, it should become possible in the near future also to measure molecules related to global warming such as CH_4, CO_2, CO and N_2O.

N_2O is the only air pollutant that has an absorption band in the visible part of the spectrum. One example of a DIAL measurement of the altitude distribution of N_2O density is shown in figure 9.7. In this experiment, a Nd:YAG pumped dye laser was used to generate radiation at 488 nm and 446.6 nm. The laser was operated at energy of 20 mJ and repetition rate of 10 Hz. The diameter of the telescope was 0.5 m, the detection range

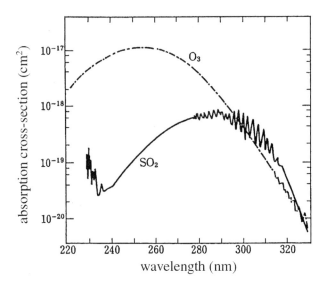

Figure 9.8. Absorption cross-sections in the ultraviolet for SO_2 and O_3 [10].

corresponded to an altitude of 1.5 km, the distance resolution was 100 m and the sensitivity of the system was estimated at 2 ppb.

In figure 9.8, the UV absorption spectra of SO_2 and O_3 are shown [10]. The SO_2 absorption band below 300 nm overlaps with an O_3 absorption band, and in real measurements care must be taken to separate contributions from each molecule. Over short distances, however, SO_2 and O_3 can be measured with the same degree of sensitivity as NO_2.

The measurement of O_3 is perhaps the best example of the effectiveness of the DIAL technique [10]. The ozone distribution has a peak at an altitude of just over 20 km, and the density is about an order of magnitude smaller at the Earth's surface. In the ultraviolet absorption band (figure 9.8), there is virtually no structure and so, in measurements, accurate wavelength tuning of the laser is not necessary. However, the choice of ON and OFF wavelengths must be made carefully. If the ON and OFF wavelengths are close together, the difference between the absorption cross-sections is small, which limits the accuracy of the method. If the wavelengths are separated by too much, though, the correction for the wavelength dependence of Mie scattering becomes important and may result in errors. In general, if the ON wavelength is chosen to be short, the absorption cross-section is large, and low altitudes can be measured accurately; if a longer wavelength is chosen, the measurable distance increases and higher altitudes can be measured. Hence, by varying the wavelength of the transmitted laser beam with respect to altitude, data with a high degree of accuracy over a wide range of altitude can be obtained.

LIDAR measurement of ozone using the DIAL technique was first carried out using a dye laser, but since then, high-power excimer lasers combined with

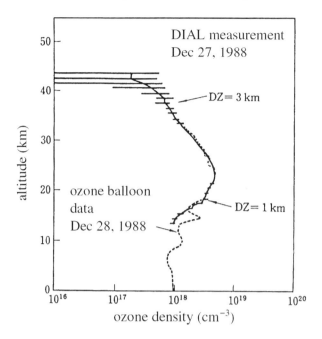

Figure 9.9. Distribution of ozone in the atmosphere measured using the ozone LIDAR system at the Meteorological Research Institute (MRI), Japan [11].

Raman-shifting devices have been used as the radiation sources, and the altitude distribution of ozone has been measured up to about 50 km. The wavelengths that are most commonly used are the 308 nm line from a XeCl excimer laser; 277 nm, obtained from the first Stokes line produced by a KrF laser in a H_2 Raman cell; 313 nm, the second Stokes line from the same system; and 291 nm, from the second Stokes line from the same laser combined with a D_2 Raman cell. Figure 9.9 shows the ozone LIDAR instrument of the Meteorological Research Institute in Japan. It uses a XeCl laser and the third-harmonic beam of a Nd:YAG laser [11]. In addition to these kinds of systems, ozone measurements also are being attempted using infrared absorption lines.

The destruction of ozone in the stratosphere and the generation of a hole in the ozone layer over the South Pole have become significant environmental concerns. DIAL instruments based on ultraviolet wavelengths are the main means by which remote sensing of the ozone distribution is conducted, and each region in the world has constructed its own measurement system. In addition, LIDAR systems mounted in aeroplanes and ships also are being tested [12]. Amongst the various different systems, the largest scale LIDAR instrument is that of the Japanese National Environmental Research Institute. This system is shown in figure 9.10. It uses a combination of three kinds of excimer lasers (KrF, XeF and XeCl) and H_2 and D_2 Raman-shifting cells to generate six wavelengths which can be simultaneously transmitted and received. Two

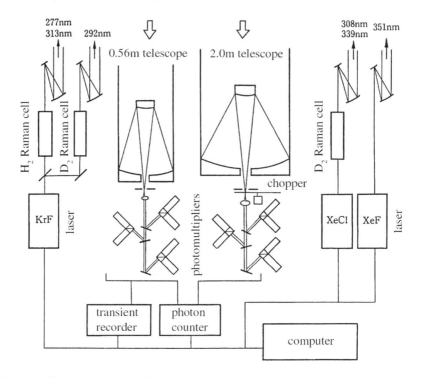

Figure 9.10. Arrangement of the large-scale ozone LIDAR system at the National Institute of Environmental Studies (NIES), Japan [13].

telescopes, with 2 m and 0.5 m diameters, are used to detect the radiation [13].

The above discussion has centred mainly on measurements using UV transitions, but most molecules in the atmosphere also have many absorption lines in the infrared. Of the many theoretically possible measurements, however, only the measurement of H_2O (i.e. the measurement of humidity) has been developed into a practical technique. The H_2O molecule has many weak transitions at wavelengths longer than about 700 nm, and measurements using sources such as dye lasers and tunable ruby lasers have been made. The recently developed alexandrite and Ti-sapphire lasers are tunable in the 700 nm region, and are very suitable for LIDAR measurements. These laser sources provide previously unavailable high-power tunable infrared radiation.

Figure 9.11 shows a H_2O density profile obtained by a DIAL system using a flashlamp-pumped, Q-switched alexandrite laser. Also shown is the H_2O density profile measured using a conventional balloon-based device. The ON wavelength for the DIAL measurement was 725.59 nm, the laser output was about 50 mJ and it was operated at 25 Hz. With this instrument, it was possible to measure the distribution to an altitude of about 10 km [14].

For infrared DIAL measurements, there have been many attempts to utilize

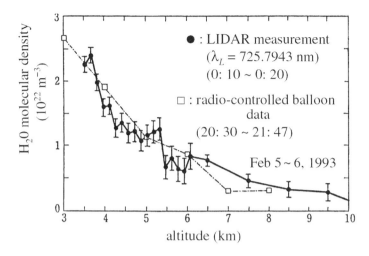

Figure 9.11. H_2O density distribution in the atmosphere, measured by a DIAL system using an alexandrite laser (measurement made at Fukuoka, Japan)

laser lines of the CO_2 laser which happen to coincide with absorption lines of particular molecules. As explained above, however, tunable solid-state lasers and optical parametric oscillators are now capable of producing good quality beams with output energies of tens of millijoules. From computer simulations, it seems that measurements of molecules associated with global warming, such as CH_4 (~ 2.2 µm), CO_2 (~ 2.0 µm), CO (~ 4.7 µm) and N_2O (~ 3.9 µm), up to altitudes of several kilometres now can be considered possible [15].

9.2.4 Raman LIDAR

The basic principle behind Raman LIDAR is that the beam from a fixed wavelength, solid-state laser is sent into the atmosphere and a variety of molecules are detected simultaneously by measuring the spectrum of the scattered radiation. However, Raman scattering cross-sections are extremely small, and measurements of most molecules are not really practical [16]. The recent development of high-power lasers operating at short wavelengths, for which the Raman scattering cross-sections become larger, has meant that measurement of molecules such as CO_2 and H_2O has become possible, at least over short distances. In particular, it has become possible to measure H_2O for altitudes up to 5 km, using the KrF laser (249 nm) or the fourth-harmonic beam of the Nd:YAG laser (266 nm) [17]. In detection sensitivity, this measurement method is comparable with DIAL methods.

It is possible to obtain Raman signals from N_2, the main constituent of the atmosphere, up to altitudes of 35 km because, even though the scattering cross-section is small, the N_2 density is very high. The atmospheric temperature then can be determined using the analysis described above for Rayleigh scattering LIDAR [18]. Rayleigh scattering LIDAR cannot be used to measure altitudes of less than about 30 km because of the interference from aerosols, but Raman LIDAR measurements are possible up to this altitude. Hence, by combining Rayleigh scattering LIDAR with Raman LIDAR, the temperature distribution at all altitudes can be measured.

Other methods have been proposed for the measurement of temperature by Raman scattering, such as using the change in the rotational Raman spectrum of molecules, or comparing the intensity of the anti-Stokes lines. However, these methods have yet to be developed to a practical level.

REFERENCES

[9.1] Hinkley E D ed 1976 *Laser Monitoring of the Atmosphere* (Spring-Verlag)
[9.2] Kobayashi T 1987 *Remote Sensing Reviews* **3** 1
[9.3] Asai K *et al* 1989 *Laser Kenkyu* **17** 292 (in Japanese)
[9.4] McCormik M P 1996 *Advances in Atmospheric Remote Sensing with Lidar* (Springer) p 141
[9.5] Takeuchi N *et al* 1983 *Ohyo Butsuri* **52** 644 (in Japanese)
[9.6] Sasano Y *et al* 1982 *J. Meteor. Sect. Japan* **60** 889
[9.7] Hirono M *et al* 1984 *Geof. Int.* **23-2** 259
[9.8] Chanin M L and Hauchecorne A 1981 *J. Geophys. Res.* **86** 9715
[9.9] Shibata T, Kobuchi M and Maeda M 1986 *Appl. Opt.* **25** 685
[9.10] Uchino O, Maeda M and Hirono M 1979 *IEEE J. Quantum Electronics* **QE-15** 1094
[9.11] Uchino O and Tabata I 1991 *Appl. Opt.* **30** 2005
[9.12] Browell E V 1989 *Proc. IEEE* **77** 419
[9.13] Sugimoto N 1993 *Opto-electronics* **8** 161
[9.14] Uchiumi M *et al* 1993 *Laser Kenkyu* **21** 1031 (in Japanese)
[9.15] Uchiumi M and Maeda M 1994 *Laser Kenkyu* **22** 448 (in Japanese)
[9.16] Inaba H and Kobayashi T 1972 *Opto-Electronics* **4** 101
[9.17] Renaut D and Capitini R 1988 *J. Atmospheric Oceanic Tech.* **5** 585
[9.18] Keckhut P, Chanin M L and Hauchecorne A 1990 *Appl. Opt.* **29** 5182

Index

Abel inversion   42, 122
ablation   233
absorption   32
absorption cross-section   36
active mode-locking   69
active plasma spectroscopy   117
active sensing   20
aerosol   4, 281
alexandrite laser   81
Allis diagram   34
analytical chemistry   258
anisotropy in electron temperature   172
anti-Stokes component   260
anti-Stokes line   89
anti-Stokes Raman scattering   31
AO (acousto-optic) effect   102
AR (Anti-reflection) coating   103
arc plasma processing   7
ArF laser   71, 74
Ar-ion laser   71, 74
ASE (amplified spontaneous emission)   240
atmosphere   4, 276
avalanche photodiode   107
axial flow fan   221
axial mode   64

Babinet-Soleil compensator   103
BBO   86
beam dump   127
beam quality   69
bolometer   120
Boltzmann distribution   195
boxcar integrator   111, 143

Bremsstrahlung   120

CARS (coherent anti-Stokes Raman spectroscopy)   191, 201
cavity ring down spectroscopy (CRDS)   38
CCD (charge-coupled device)   109, 254
certirfugal turbo-compressor   219
Child-Langmuir's formula   149
classical electron radius   29
CO_2 laser   72, 230
coherence   14
coherence length   16
coherent anti-Stoles Raman Spectroscopy (CARS)   191, 201
collective Thomson scattering   47
collinear arrangement   202
collisional narrowing   202
colour-schlieren   211
combustion   5, 188
combustion measurement   188
Compact Helical System (CHS)   131
compressor   4
computer tomography (CT)   43
condensed matter   3, 13
controlled thermonuclear fusion   7
correlation reflectometer   124
crossed-beam arrangement   201
CT (computer tomography)   43
cut-off   34
Cu-vapour laser   70, 234
CVD (chemical vapour deposition)   233
cyclotron emission   120
cyclotron frequency   11

cyclotron motion 11
Czerny-Turner spectrometer 92

DC arc discharge in atmosphere 162
Debye length 10, 46
Debye shielding distance 10
de-convolution 58
degenerative four-wave mixing (DFWM) 191, 193, 204
detonation wave 211
DFB (distributed feedback) 78
DFWM (degenerative four-wave mixing) 191, 193, 204
DIAL (differential absorption LIDAR) 279, 285
Differential absorption LIDAR (DIAL) 279, 285
differential scattering cross section 27
diffraction 15
diffraction grating 66, 90, 91
diffraction limit 19
discharge-pumped excimer laser 164
dispersion relation 34
dispersivity 91
doping 233
Doppler broadening 55, 61
Doppler profile 194, 236
Doppler-broadened profile 248
Doppler-free spectroscopy 22, 258
drift motion 11
dye laser 79

echellette grating 91
ECR (electron cyclotron resonance) 38
ECRH (electron cyclotron resonance heating) 13
Einstein coefficients 32, 60
elastic scattering 29
electric field 136, 139
electron cyclotron resonance (ECR) plasma 38, 169
electron cyclotron resonance heating (ECRH) 13
electron temperature 9
electro-optic (EO) effect 99
etalon 66, 95
excimer laser 74, 231
excitation spectrum 22, 58

Fabry-Perot interferometer 95

Fabry-Perot resonator 63
fan 4
Faraday rotation method 41
fine structure broadening 55
finesse 96
flashlamp pumping 76
flow velocity 214
fluctuation 9
fluid dynamic turbulence 9
fluorescence 32, 50, 61
fluorescence (resonant) scattering 50, 131
fluorescence lifetime 61
fluorescence spectroscopy 32
four states of matter 13
Fourier spectroscopy 95
Fraunhofer diffraction method 136
Fraunhofer's diffraction theory 18
free energy 9
free spectral range (FSR) 96
frequency mixing 85
Fresnel rhomb 103
fringe 39
FSR (free spectral range) 96
FT-IR (Fourier transform infrared) spectroscopy 95, 191
fuel 5

GaAs 77
gas laser 70
Gaussian beam 65
graded-index fibre 97
gyrotron 130

H atom 197
H_2O 288
half-waveplate 103
halo 121
Heliotron E 136
He-Ne laser 70
high resolution 20
higher harmonic generation 85
holography 213
homogeneous broadening 55
hot band 56
hydrocarbon 5

identification of spectral line 118
idler 85
image intensifier (II) 109
image technique 189

imaging   23, 252
in situ   237
incoherent   14
incoherent Thomson scattering   46
induction coupled plasma (ICP)   174
infrared absorption spectroscopy (IRAS)   191
infrared laser   72
inhomogeneous broadening   55
interference (IF)   15, 214
interferometer   16, 93
interferometry   38, 121, 212
intra-cavity absoption   38
ion laser   71
ion temperature   9
ionized gas   6
IRAS (infrared absorption spectroscopy)   191, 238
isotope separation   234

JET (Joint European Torus) tokamak   127, 138
Johnson noise   106

KDP   86
Kerr effect   99
KrF laser   74
Kr-ion laser   71
KTP   86

Larmor radius   11
laser   60
laser 2 focus (L2F) method   214
laser breakdown spectroscopy   263
laser Doppler velocimeter (LDV)   196, 215, 217, 220
laser induced emission spectroscopy   262
laser interferometer   40, 154
laser ionization spectroscopy   268
laser mass spectroscopy (LMS)   53
laser microscopy   23, 262
laser optogalvanic (LOG) method   139, 141
laser phase contrast method   136
laser processing   230
laser pumping   7
laser radar   25, 276
laser Raman spectroscopy   260
laser spectroscopy   22, 258

laser spectrum   66
laser-enhanced ionization (LEI) spectroscopy   271
laser-induced collisional fluorescence (LICF)   141, 147
laser-induced fluorescence (LIF)   131, 191, 222, 237, 264
Laudau damping   47
LD (laser diode)   77
LD-pumped solid state laser   79
LDV (laser Doppler velocimeter)   196, 215, 217, 220
LEI (laser-enhanced ionization)   271
LICF (laser-induced collisional fluorescence)   141, 147
LIDAR (light detection and ranging)   25, 276
LIDAR Thomson scattering   127
LIF (laser induced fluorescence)   50, 131, 141, 191, 222, 237, 238, 241, 264
light detector   103
light source   7
$LiNbO_3$ 86
lock-in amplifier   110
longitudinal mode   64

Mach-Zehnder interferometer   39, 95, 121, 212
magnetic field   136
magnetohydrodynamic (MHD) power generation   7
magnetron discharge   180
material fabrication   7
measurement beam   39
MHD (magnetohydrodynamic) power generation   7
Michelson interferometer   16, 39, 93, 121
micro-lithography   233
Mie Scattering   29, 48, 189, 280
Mie scattering LIDAR   280
MO (magneto-optic)effect   101
mode   64
mode-locking   68
multipass absorption   37
multi-photon process   85

N_2 laser   75
NA (numerical aperture)   97
natural broadening   54, 61

natural width 54, 61
Nd:YAG laser 76, 234
NEP (noise-equivalent power) 107
non-condensed matter 3, 13
nonlinear optical effect 24, 82
nonlinear optics 24
nonlinear Raman spectroscopy 49
non-perturbing measurement 24
non-radiative transition 60
normal shock wave 208
numerical aperture (NA) 97

OH radical 200
OMA (optical multichannel analyzer) 109
O-mode 34, 40
OPO (optical parametric oscillator) 85, 87
optical component 100
optical parametric amplification 85
optical waveguide 97
optogalvanic spectroscopy (OGS) 269
ozone 286

parametric effect 83
PAS (photoacoustic spectroscopy) 272
passive mode-locking 69
Pb-calcogenide laser 78
phase (or index) matching condition 83, 192
phase conjugation 193
phase matching 83, 192
phase-sensitive detector (PSD) 110
phonon-terminated laser 81
photo-acoustic spectroscopy 32, 272
photoconduction cell 107
photodiode 106
photo-etching 233
photo-ionization 33, 52
photomultiplier tube (PMT) 104
photon counter 111
photothermal spectroscopy 271
phototube 104
PIN photodiode 107
plane wave 17
plasma 6
plasma chemistry 7
plasma display panel 7
plasma frequency 10
plasma oscillation 10
plasma spectroscopy 117

PLD (pulsed laser deposition) 233, 247
plume 231
Pockels effect 99
polarimetry 41
polarization 28
polarization conserving fibre 98
population inversion 63
power density 19
pressure broadening 62
pressure equipment 5
pressure-broadened width 54
prism 66, 90
propulsion in space 7
PSD (phase-sensitive detector) 110
pulse duration 67
pulsed arc discharge in atmosphere 155, 159

Q factor 63
Q-switching 67
quarter-waveplate 103

radiative transition 60
radical 8
Raman laser 85
Raman LIDAR 279, 289
Raman scattering 31, 48, 190, 289
Raman shifter 89
Raman spectroscopy 31
Rankine-Hugonoit relation 212
rapid-frequency-scan (RAFS) laser 135
Rayleigh Scattering 29, 48, 189, 193, 283
Rayleigh scattering LIDAR 283
reference beam 39
reflection 26, 33
reflecto-interferometer 36, 123
reflectrometer 36, 123, 136
refraction 26, 43
refractive index change 208
rejection 167
remote sensing 24, 276
resonance ionization spectroscopy (RIS) 52, 269
resonant absorption 32
resonant scattering 50
RIS (Resonant ionization spectroscopy) 245, 269
rotational transition 73
Rowland circle 92

ruby laser 76
Rydberg level 140

scattering 26, 45
scattering angle 29
schlieren method 43, 209
second harmonic generation (SHG) 83
semiconductor laser 77
semiconductor processing 7
shadowgraphy 43, 45, 211
SHG (second harmonic generation) 85
shock tube 211
shock wave 208
short pulse 19
shot noise 106
signal 84
signal wave 84, 87
signal-to-noise ratio (S/N) 21
single mode 67
single pass absorption 37
size parameter 29
solid-state laser 75
spatial coherence 17
spatial resolution 22
spectral line intensity 118
spectral line shape 119
spectral profile measurement 54
spectral resolution 22
spectrometer 90, 92
spherial wave 17
spontaneous emission 14, 32
sputter deposition 179
Stark broadening 54
Stark effect 54, 139
Stark mixing 140, 150
Stark shifting 140
Stark splitting 140
step-index fiber 97
stepper 233
stimulated Brillouin scattering 88
stimulated emission 14
stimulated Raman scattering 88
Stokes component 260
Stokes line 89
Stokes Raman scattering 31
stray (laser) light 51, 125
streak camera 112
surface modification 7
synchronous pumping 69
synchronous scan luminescence 265

T-3 tokamak 125
TEA (transversely-excited atmospheric) laser 73
temperature measurement 193, 251, 284
temporal coherence 15
temporal resolution 22
TEXTOR tokamak 121, 133, 137
thermal detector 108
thermal fluctuation 9
thermal lens spectroscopy 32, 273
thermal treatment process 288
thermal velocity 11
thermal-light-diffraction method 274
Thompson distribution 183
Thomson scattering 28, 46, 124, 159
three states of matter 3
threshold 63
Ti:sapphire laser 81
time-of-flight (TOF) 236, 249
total radiation density 120
transmission 26, 36
transonic axial flow fan 222
transport phenomena 9
transverse mode 65
tunable laser 22, 66, 79, 81
tunable solid-state laser 81
turbulence 9
TVTS (television Thomson scattering) system 126
Twyman-Green interferometer 95

ultraviolet laser 73
unstable resonator 69

vacuum equipment 5
various states of matter 13
velocity measurement 196
Verdet's constant 101
vibrational transition 74
viewing dump 127

wavefront 17
White cell 38

XeCl laser 74

YBCO 247

Zeeman broadening 55

RETURN TO ➡	PHYSICS LIBRARY 351 LeConte Hall		642-3122
LOAN PERIOD 1 **1-MONTH**	2		3
4	5		6

ALL BOOKS MAY BE RECALLED AFTER 7 DAYS
Overdue books are subject to replacement bills

DUE AS STAMPED BELOW

This book will be held
in PHYSICS LIBRARY
until **FEB 2 6 2002**

MAY 20 1 2004
MAY 2 4 2004

MAY 2 1 2006

UNIVERSITY OF CALIFORNIA, BERKELEY
BERKELEY, CA 94720

FORM NO. DD 25